Applied Industrial Catalysis
Volume 1

Contents of Other Volumes

Applied Industrial Catalysis

Volume 1

Edited by

BRUCE E. LEACH

Conoco Inc.
Research and Development
Ponca City, Oklahoma

1983

ACADEMIC PRESS

A Subsidiary of Harcourt Brace Jovanovich, Publishers

New York   London

Paris   San Diego   San Francisco   São Paulo   Sydney   Tokyo   Toronto

ACADEMIC PRESS, INC.
111 Fifth Avenue, New York, New York 10003

United Kingdom Edition published by
ACADEMIC PRESS, INC. (LONDON) LTD.
24/28 Oval Road, London NW1 7DX

Library of Congress Cataloging in Publication Data

Main entry under title:

Applied industrial catalysis.

Includes index.
1. Catalysis. I. Leach, Bruce E.
TP156.C35A66 1983   660.2'995   82-22751
ISBN 0-12-440201-1 (v. 1)

PRINTED IN THE UNITED STATES OF AMERICA

83 84 85 86   9 8 7 6 5 4 3 2 1

Contents

4 Catalytic Hydrotreating in Petroleum Refining

D. C. McCulloch

5 Catalytic Reforming of Naphtha in Petroleum Refineries

M. D. Edgar

6 Catalysis of the Phillips Petroleum Company Polyethylene Process

J. P. Hogan

7 The Evolution of Ziegler–Natta Catalysts for Propylene Polymerization

K. B. Triplett

8 Ethylene Oxide Synthesis

J. M. BERTY

9 Oxychlorination of Ethylene

J. S. NAWORSKI AND E. S. VELEZ

10 Methanol Carbonylation to Acetic Acid Processes

R. T. EBY AND T. C. SINGLETON

Contributors

Numbers in parentheses indicate the pages on which the authors' contributions begin.

J. M. BERTY* (43, 207), Berty Reaction Engineers, Ltd., Erie, Pennsylvania 16508

R. T. EBY (275), Process Technology Department, Monsanto Company, Texas City, Texas 77590

M. D. EDGAR (123), Catalyst Department, American Cyanamid Company, Houston, Texas 77001

J. P. HOGAN (149), Research Center, Phillips Petroleum Company, Bartlesville, Oklahoma 74003

B. E. LEACH (1), Research and Development Center, Conoco, Inc., Ponca City, Oklahoma 74601

D. C. MCCULLOCH (69), Worldwide Catalyst Department, American Cyanamid Company, Wayne, New Jersey 07470

J. S. NAWORSKI (239), Richmond Research Center, Stauffer Chemical Company, Richmond, California 94804

E. F. SANDERS (31), Mallinckrodt Inc., Calsicat Division, Erie, Pennsylvania 16503

E. J. SCHLOSSMACHER† (31), Mallinckrodt Inc., Calsicat Division, Erie, Pennsylvania 16503

T. C. SINGLETON (275), Process Technology Department, Monsanto Company, Texas City, Texas 77590

K. B. TRIPLETT (177), Specialty Chemical Division, Stauffer Chemical Company, Dobbs Ferry, New York 10522

E. S. VELEZ (239), Richmond Research Center, Stauffer Chemical Company, Richmond, California 94804

*Present address: Chemical Engineering, University of Akron, Akron, Ohio 44325.

†Present address: Research and Development Department, Ashland Petroleum Company, Ashland, Kentucky 41101.

Preface

Industrial catalysis contributes significantly to our modern economy and life-style. The objective of these volumes is a practical description of catalysis by industrial scientists. Excellent reference works on catalyst theory, kinetics, and reaction mechanisms already exist, but reviews of practical operation of commercial units are rarely published by industrial scientists.

Industrial catalysis is influenced not only by science but also by business, economics, markets, and politics. These factors are discussed in various chapters. The reader should recognize that in most cases competitive technology exists for the synthesis of chemical intermediates. The examples given represent current industrial practice but obviously do not disclose proprietary information.

This first of three volumes begins with a review of the importance of industrial catalysis and its effect on our life-style and environment (Chapter 1, by B. E. Leach). In Chapter 2, E. F. Sanders and E. J. Schlossmacher describe how to take a laboratory catalyst to successful commercialization with a minimum of problems. In Chapter 3, J. M. Berty presents techniques for evaluating a catalyst in laboratory reactors. In Chapters 4 and 5, D. C. McCulloch and M. D. Edgar, respectively, describe in detail two major refinery processes—hydrotreating and reforming. In Chapters 6 and 7, specific processes for polyethylene (J. P. Hogan) and polypropylene (K. B. Triplett) manufacture are reviewed. Ethylene oxide synthesis is described in Chapter 8 by J. M. Berty. Oxychlorination of ethylene to ethylene dichloride is the subject of Chapter 9 by J. S. Naworski and E. S. Velez. Methanol carbonylation to acetic acid is reviewed by R. T. Eby and T. C. Singleton in Chapter 10. Additional general catalysis and specific processes will be presented in chapters to be published in subsequent volumes.

The editor acknowledges Drs. D. P. Higley, R. J. Convers, and M. L. Shannon, who assisted in reviewing chapters. Their critical comments and helpful suggestions are appreciated. Dr. C. M. Starks, who encouraged me to edit this work, and Sherry Martin, who assisted with communications with the authors, deserve special recognition. I also wish to thank my wife, Sharon, who supported me in the many extra hours of work necessary for this project.

It is the editor's hope that this book will be of value to all those active in catalysis and also that it will promote an understanding of industrial catalyst research and processes that could be valuable to seniors and graduate students preparing for industrial careers.

CHAPTER 1

Industrial Catalysis: Chemistry Applied to Your Life-Style and Environment

BRUCE E. LEACH

Research and Development Department
Conoco, Inc.
Ponca City, Oklahoma

I. Industrial Catalysis: Definition, Scope, and Importance

Catalysis is a major factor in industrial research, process selection, plant design, and plant operation. The success of the chemical industry is based largely on catalyst technology. The discovery of new catalysts and their application have historically led to major innovations in chemical processing. Market, business, and political factors combine to encourage or require further improvements in catalyst technology with time.

It is the interaction of business, markets, economics, and politics with chemistry that distinguishes industry from academia. Chemical principles and the laws of thermodynamics still apply. The definition of a catalyst is the same—a material that changes the rate of a reaction without itself being consumed. Catalysts have no effect on the position of equilibrium, and one cannot make a reaction proceed that is forbidden by the laws of thermodynamics. A catalyst acts to lower the activation energy barrier for reactions that have a net decrease in free energy. The alternate reaction paths provided by catalysts within the laws of thermodynamics and chemistry add value to feedstock materials in the refining and chemical processing industry.

The fundamental aspects of adsorption on active sites are adequately covered in most catalysis reference works. The catalysis scientist should be knowledgeable about adsorption and kinetic experiments and their results. For example, in the reaction $A \rightarrow B$ (Fig. 1), a series of steps in the catalytic reaction can be considered: (1) external diffusion, $k_{\text{ext. diff.}} = k_g s_{\text{ext}}$; (2) internal diffusion, $k_{\text{int. diff.}} = k_s S_{\text{int}} \eta$; (3) adsorption, k_{ads}; (4) surface reaction, k_s; (5) desorption of B; (6) internal diffusion of B; and (7) external diffusion of B.

In external diffusion the size and shape of the catalyst particle, the volume/diameter ratio of the reactor, and the space velocity are factors that influence the amount of channeling, the type of flow, the extent of back-mixing, and the residence time. Guidelines for catalyst evaluations in laboratory reactors are given in Chapter 3.

Internal diffusion is dependent on the pore structure of the catalyst. Pores are arbitrarily placed in three size categories according to pore diameter: micropores, <15 Å; mesopores, $15-150$ Å; macropores, >150 Å. The pore diameter affects internal diffusion. The surface area of many catalysts is primarily internal surface, so most interactions and collisions occur with the internal surface. Pores have a variety of structures, and there is an effectiveness factor in the equation expressing internal diffusion rates that depends on how difficult it is for reactants to diffuse in or out. Often it is not so critical to know the value of k for internal diffusion as to know whether the reaction occurs primarily on the external surface or within the pores of the catalyst and to recognize the consequences. For example, in a consecutive reaction

$$A + B \rightarrow C$$
$$C + B \rightarrow D$$

where C is the desired product and D is to be minimized, there are two recognized methods for achieving the desired result. The pore size can be kept small enough so that the reaction occurs primarily on the surface of the catalyst particles. Examples include most hydrocarbon oxidation catalysts where a decreased rate of reaction due to reduced surface area (internal) is

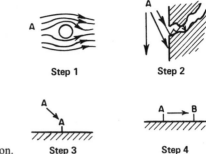

Fig. 1. Steps in the catalytic reaction.   Step 3   Step 4

acceptable to gain selectivity of product. Specific examples are given in Chapter 8.

Alternatively, a large pore size can be selected so that there is facile diffusion out of the pore system before C can react further with B. This approach reduces the internal surface area, whereas in the case of very small pores the internal surface is inaccessible.

Another example of pore size importance is found in the hydrodesulfurization of heavy crude oils. Rapid catalyst deactivation is associated with particular pore sizes, and approaches based on supports of either small [1] or large [2] pore diameter have been developed.

Theory has often come after discovery. Ideally, the objective is to design a catalyst based on first principles. This objective is not yet within the capability of catalysis science. In addition, the complicating factor of economics must be addressed in the industrial setting.

Optimization of the added value to feedstocks requires a knowledge and interaction of chemistry and economics. This is a special challenge to the scientist or engineer in industrial catalysis. Both catalytic science and economics change with time. This ensures that new and improved catalysts will continue to be developed. These catalysts will be used to process historic and new feedstocks. The variables encompassed by new catalysts, new and modified processes, and the changing economics of alternate feedstocks taken together with political and environmental restraints make for many exciting and challenging technical endeavors within the industry.

The scope of catalysis in industry ranges from theoretical predictions of catalytic activity to the art of catalyst forming. It includes both work on the frontiers of catalysis science and the careful recommendation of a particular commercial catalyst that meets a specific customer's feedstock and reactor design criteria. It includes catalyst regeneration, testing, and quality control, as well as catalyst selection. Finding and removing a catalyst poison present in a part per million or even a part per billion quantity in the feedstock presents its own set of analytical and chemical challenges.

A major change in the practice of catalysis in the past several decades has been the concerted application of analytical techniques to catalysts. A number of these techniques and their application in catalysis are given in Table I.

Industrial catalysis provides an opportunity for technical exchange with many disciplines. Modern catalysts must rely on the integration of a broad range of technical expertise and experimental capabilities. Chemical engineering, organic chemistry, inorganic chemistry, coordination chemistry, analytical chemistry, and surface science are all essential in understanding and developing catalysts (Fig. 2).

At a time at which the conversion to SI units is in progress and with older

TABLE I

Analytical Techniques Frequently Used in Catalysis

Characterization	Technique
Bulk	Surface area analysis
	Pore volume analysis
	X rays
	Scanning electron microscope
	Electron microprobe
	Infrared spectroscopy
	Elemental analysis
	Surface acidity
	Loss on ignition
	Thermal gravimetric analysis
	Density
	Bulk crushing strength
	Particle size analysis
	NMR
	ESR
Surface	X-ray scattering
	Laser Raman spectroscopy
	Extended x-ray adsorption fine-structure microscopy (EXAFS)
	X-ray photoelectron spectroscopy (ESCA)
	Scanning electron microscope (SEM)
	Auger electron spectroscopy (AES)
	Ion-scattering spectroscopy (ISS)
	Secondary ion mass spectroscopy (SIMS)
	Magnetic susceptibility
	Selective surface area
	Selective adsorption
	Programmed temperature desorption

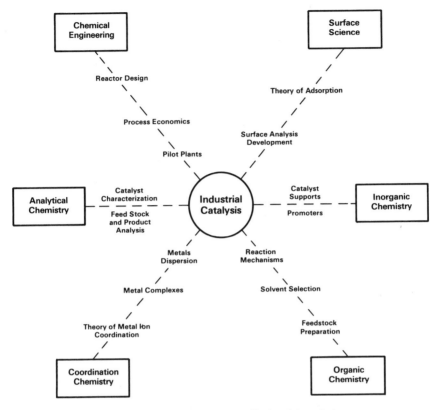

Fig. 2. Technical dependence of industrial catalysis.

units still widely used, there is a communication problem. This book reflects problems that will continue for many years. For this reason a common nomenclature and units table is provided in the Appendix to this volume. Although the reader may find it confusing to see temperature in degrees centigrade, degrees Farenheit, and kelvins all in the same chapter, this is the reality of the situation one is faced with in industry.

The scope of catalysts in the chemicals industry is so extensive that it is rare to find a research problem that cannot be redefined in terms of catalysis. Thus although the research problem title and the objective may be stated in business and economic language, the scientific methodology required for achieving the project goals often involves the practice of catalytic science. The major innovations in the petrochemical industry in the past 25 yr have involved breakthroughs in catalyst research. In most industrial research problems, catalyst selection or improvement is the key to the success of the project. This heavy dependence on catalysis can be seen in both exploratory and applied research.

The shutdown of a chemical plant because of catalyst problems is a crisis situation. This subject will be developed later in the chapter. Timely action is possible only if the catalysis scientist has the background and expertise to diagnose and prescribe a remedy. Considerable effort is devoted in industry to evaluating changes in catalysts, feedstocks, and process conditions (including upset conditions) and their effect on catalyst lifetime, selectivity, and productivity. Losses from the shutdown of large chemical plants can be many hundred thousand dollars per day. With this much at stake, there is justification for considerable research. Operating departments often support extensive piloting facilities after plant start-up as well as before plant construction to provide insurance that the downtime at a large facility will be minimal. The speedy resolution of catalyst-related problems at operating facilities may not lead to publications, patents, and recognition from the scientific community, but there is a sense of personal accomplishment and recognition from catalyst marketing managers, chemical plant managers, and research directors. It is the profitability of current chemical plants that justifies future expansion and research in industrial catalysis.

II.   History of Industrial Catalysis

A catalyst was used industrially for the first time by J. Roebuck in the manufacture of lead chamber sulfuric acid in 1746. At that time Berzelius had not yet used the word "catalysis" — that came in 1836. Early development occurred in inorganic industrial chemistry with processes for carbon dioxide, sulfur trioxide, and chlorine production in the 1800s. P. Sabatier and R. Senderen in 1897 found that nickel was a good hydrogenation catalyst. P. Sabatier [3], in his book *Catalysis in Organic Chemistry,* gives an excellent perspective of catalysis in the early 1900s. It was a time when answers to questions about transition states, adsorption, and mechanisms were difficult to obtain, and yet Sabatier was asking the right questions. His idea of temporary unstable intermediate compounds being formed in catalysis was correct. He lamented the unsatisfactory state of knowledge, yet the period 1900–1920 saw advances in many areas. It was the time of Ostwald, Gibbs, Bosch, Ipatief, Einstein, Planck, Bohr, and Rutherford, among others. Scientists such as E. Fischer, Kekulé, Claisen, Fittig, Sandmeyer, Faworsky, Deacon, Dewar, Friedel, and Crafts had made their contributions to organic chemistry just prior to 1900.

Initially most catalysts were relatively pure compounds. Multicomponent catalysts were studied after 1900 at Badische Anilin- & Soda-Fabrik (BASF). Haber discovered ammonia synthesis at high pressure using osmium or uranium catalysts. Bosch and associates at BASF developed the use of magnetite promoted with alumina and alkali. This research project is described by A. Mittasch [4] in detail.

The ammonia synthesis industry is based on promoted iron catalysts. A catalyst was developed to supply hydrogen via the water gas shift reaction by the BASF group.

Bosch at BASF next attempted to reduce carbon monoxide with hydrogen at high pressures to produce alcohols and higher hydrocarbons. This work led to methanol synthesis using alkali-promoted zinc oxide plus chromium oxide in 1923. Synthetic hydrocarbons were made from synthesis gas in 1927 by Fischer and Tropsch.

The adsorption of reactants on catalyst surfaces was first thought to be important in the 1900–1920 period. Langmuir–Hinshelwood and Rideal–Eley mechanisms were proposed. The adsorption of gases by solids and particularly the adsorption of hydrogen presented many unknowns. For example, it was not known why the quantity of hydrogen adsorbed varied or indeed how a substance like palladium could adsorb so much hydrogen.

A major development in the 1920s occurred when H. S. Taylor distinguished among activated adsorption, chemisorption, and physical adsorption. He also developed the concept of active centers.

In the 1930s a number of advances occurred that aided in the study of adsorption:

(1) Isotopes became available in 1933.

(2) Brunauer and Emmett discovered how to measure the surface area and pore geometry of catalysts using physical adsorption.

(3) Beeck used evaporated metal films for basic catalytic studies.

(4) Roberts made tungsten filaments—for the first time "clean" surfaces could be studied because tungsten could be heated hot enough to clean metal surfaces.

(5) Rideal made other metal filaments and films.

Adsorption studies dominated catalysis science for a time while the new techniques were being applied. In the 1950s attention shifted to the nature of the interactions between the active center and the adsorbate, and today spectroscopic methods continue to reveal information about bonding in catalysts.

A survey of catalytic development is given in Fig. 3. The past 60 yr have been very active ones in the development of new catalytic processes. A list [5] of the more recent of these has been compiled by Halcon International (Table II). Other significant catalyst developments include the family of ZSM zeolites discovered by Mobil Oil Corporation, the carbonylation of methanol to acetic acid practiced by Monsanto Company, and a new generation of catalysts for refining, polyolefins, oxychlorination, etc.

Few heterogeneous catalyst compositions remain constant for as long as a

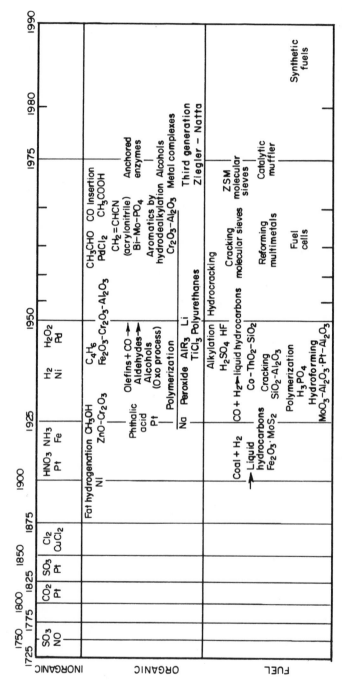

Fig. 3. Survey of catalytic development. Adapted with permission from Kirk-Othmer Encyclopedia of Chemical Technology, Third Edition, John Wiley & Sons, Inc.

8

TABLE II

Chemical Processing Developments of the Last 25 Yr as Compiled by Halcon International[a]

Approximate date	Product	Development	Company[b]			
			Chemical	Oil	Process engineering	
Before 1957	Isocyanates–urethanes	Urethanes and foams (polyether polyols, one-shot foam, etc.)	Bayer	Wyandotte	Houdry	
1953+	Ammonia	High-pressure synthetic gas	—	—	Pullman, Kellogg	
1955	Maleic anhydride	High-yield benzene oxidation	—	—	Halcon	
1958	High-density polyethylene, polypropylene	New catalysts	Montecatini	Phillips, Avisun	—	
1958	α-Olefins	New catalysts	Ethyl	Gulf, Conoco	—	
1958+	Terephthalic acid	Air oxidation of p-xylene pure product	Halcon	—	Amoco	
1959	Acetaldehyde	Vapor phase ethylene oxidation	Wacker	—	Hoechst	
1960–1970	Oxidation alcohols cyclohexanol, cyclohexanone (for nylon)	Improved catalysts	UCC, ICI	Exxon, Shell	—	
1964		Cyclohexane oxidation, boric system	—	—	Halcon	
1965+	Vinyl chloride	Oxychlorination of ethylene	Goodrich, Monsanto, Stauffer, PPG	—	—	
1965	Acrylonitrile	Propylene ammoxidation	—	SOHIO	—	
1965	HMDA (for nylon)	Acrylonitrile electrohydrodimerization	Monsanto	—	—	
1967+	Vinyl acetate	Ethylene + acetic acid + O_2, vapor phase	Bayer, Celanese, Hoechst, USI	—	—	
1968	Acetic acid	High-pressure methanol + CO	BASF, DuPont	—	—	

Table II (continues)

TABLE II (*Continued*)

Approximate date	Product	Development	Company[b] Chemical	Oil	Process engineering
1969+	Propylene oxide, glycol, TBA	Epoxidation with hydroperoxide	—	—	Arco, Halcon
1969	Phthalic anhydride	High-yield o-xylene oxid.	BASF	—	—
1969	Acrylates	Propylene oxidation	Celanese, Rohm & Haas, UCC	BP, SOHIO	—
1969	Quiana	From cyclododecane oxidation	DuPont	—	Halcon
1970+	Ethylene oxide	Catalyst improvements	Shell, UCC	—	Halcon
1970	p-Xylene	Recovery by adsorption	—	—	UOP
1970	Methanol	Low-pressure CO + H_2	ICI	—	—
1972	HMDA (for nylon)	Butadiene + HCN	DuPont	—	—
1972	Styrene and propylene oxide	Epoxidation with hydroperoxide	—	—	Arco, Halcon
1973	Acetic acid	Low-pressure methanol + CO	Monsanto	—	—
1974+	Maleic anhydride	From butane	Amoco, Monsanto	—	Halcon
1974	Kevlar	High-tensile fiber	DuPont	—	—
1974	Polypropylene	Vapor phase	BASF	—	—
1978	Ethylene glycol (and vinyl acetate)	Via acetoxylation	—	—	Halcon

[a] Reprinted with permission of the Halcon SC Group, Inc.

[b] BASF, Badische Anilin- & Sodi-Fabrik; ICI, Imperial Chemicals; PPG, Pittsburg Plate Glass; UCC, Union Carbide Corporation; UOP, Universal Oil Products.

TABLE III

Examples of Titanium-Based Polyethylene Catalyst Developments

Catalyst	Improvement[a]	Reference
$TiCl_4$ + $AlEt_3$	Basic catalyst, Ziegler and co-workers	[6]
$TiCl_3 \cdot \frac{1}{3}AlCl_3$	First-generation commercial catalyst	[7]
$TiCl_3 \cdot \frac{1}{3}AlCl_3$ + electron donor	Activity increase of ~ 10 fold	[8]
$Mg(Cl)OH$ + $TiCl_4$	Supported titanium trichloride catalyst, ~ 13.6 kg PE/hr atm ethylene g Ti	[9]
$TiCl_4$ + ($MgBuCl$ + methylhydropolysiloxane)	132 kg PE/hr g Ti	[10]

[a] PE, Polyethylene.

decade. An excellent example is provided by polyolefin catalysts based on titanium trichloride. These catalysts are now in what is called the third generation. The history of their development is briefly summarized with examples in Table III. Many permutations of each generation of catalysts exist, as evidenced by the large number of patents published by competing

TABLE IV

Some Chemicals Produced from Ethylene and Propylene

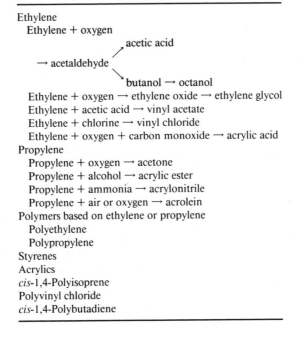

Ethylene
   Ethylene + oxygen
                   acetic acid
        $\rightarrow$ acetaldehyde
                   butanol $\rightarrow$ octanol
   Ethylene + oxygen $\rightarrow$ ethylene oxide $\rightarrow$ ethylene glycol
   Ethylene + acetic acid $\rightarrow$ vinyl acetate
   Ethylene + chlorine $\rightarrow$ vinyl chloride
   Ethylene + oxygen + carbon monoxide $\rightarrow$ acrylic acid
Propylene
   Propylene + oxygen $\rightarrow$ acetone
   Propylene + alcohol $\rightarrow$ acrylic ester
   Propylene + ammonia $\rightarrow$ acrylonitrile
   Propylene + air or oxygen $\rightarrow$ acrolein
Polymers based on ethylene or propylene
   Polyethylene
   Polypropylene
Styrenes
Acrylics
cis-1,4-Polyisoprene
Polyvinyl chloride
cis-1,4-Polybutadiene

companies. Similar examples could have been chosen in refining or for other processes.

III. Impact on Industry and Economics

The chemical industry has grown in recent decades and significantly affects the economy. Catalysts and the products derived from catalytic reactions directly or indirectly account for 10 to 15% of the gross national product (GNP) of the United States. The two major factors in the rapid development of the chemical industry since 1950 have been the low-cost petroleum-based supply of ethylene and propylene raw materials and the development of oxidation and polymerization catalysts. The diversity of chemicals produced from ethylene and propylene is illustrated in Table IV.

IV. Impact of Industrial Catalysis on Science

The beginnings of industrial catalysis were applications of basic research. The development of catalysis has been based on scientific innovation. The objective has been to design catalysts of high activity and selectivity based on scientific theory rather than trial and error. This has encouraged the development of theories explaining catalysis in terms of active sites, geometry, metal properties, etc. The objective of a complete understanding has been most elusive, and catalysis remains both an art and a science. Basic research has led to numerous models, and they have been challenged, refined, and sometimes discarded. One does not have to understand why a catalyst works to take commercial advantage of it, but it usually helps to have a fundamental understanding of the active site and the interactions of the reactants and products with the sites. Some commercial systems are so complex that they present challenges to scientific analysis that have yet to be solved even after 50 yr.

Industry and academia could both benefit by increased interaction in the catalysis area. There are many simple chemical synthesis reactions that were studied inadequately or before modern instrumentation was available, and some of these can yield interesting new aspects of chemistry. An example is phenol methylation [11] over an alumina catalyst to yield primarily o-cresol, 2,6-xylenol, and 2,3,6-trimethylphenol. It was originally believed that 2,6-xylenol first isomerized to 2,3-xylenol or 2,5-xylenol which then reacted with methanol to form 2,3,6-trimethylphenol. When the reaction was investigated under trickle bed reaction conditions at temperatures below the isomerization range, it was discovered that the selectivity for 2,3,6-trimethylphenol from 2,6-xylenol methylation increased. The reaction is now understood as occurring via an ipso mechanism [Eq. (1)].

Both industry and universities have many common objectives that further catalysis science. These include the following manifestations: training of catalysis scientists, development of centers of catalysis, publication of books and papers, consulting, catalysis meetings and conferences, seminars, research grants, and sharing of research facilities.

V. Catalysis Impact on Life-Style

A. HISTORICAL PERSPECTIVE

The impact of industrial chemistry and catalysis on life-styles during this century is dramatic. In recent times chemistry has received some negative connotations because the potential for detrimental effects on the environment had not been adequately appreciated in a few widely publicized instances. However, few people would like to do without the industrial products that have so changed civilization in the twentieth century. These impacts have been most evident in the following areas:

(1)   Transportation—fuel, tires, materials of construction (plastics), and pollution control;
(2)   Food—packaging, fertilizers, and insecticides;
(3)   Clothing—nylon, polyester, dacron, rayon, and orlon;
(4)   Detergents and cosmetics—biodegradable surfactants;
(5)   Housewares and furniture—material of construction (plastics), insulation;
(6)   Construction—carpets, plastic pipes, insulation, and engineering plastics; and
(7)   Toys—plastic construction.

Approximately 85% of organic industrial chemicals on a weight basis go into plastic applications. The role of catalysts in the preparation of monomers and/or polymers will be described in Section VII. Polymerization catalysts are the subjects of Chapter 6 in this volume and Chapter 5 in Volume 3.

B. IMPACT ON ENVIRONMENT

Industrial catalysis has responded to problems of pollution control. Catalysts are used to remove hydrocarbons, carbon monoxide, and nitrogen oxides from waste and exhaust gases. The fundamental problem is to develop chemical processes that minimize or eliminate pollution in the manufacturing process. Catalysts will play a vital role in the development of these nonpolluting processes. A major challenge will be to solve the environmental problems associated with the change in chemical feedstocks from petroleum to coal in the next century.

Catalysts themselves can pose environmental problems in manufacture and disposal. These aspects are considered in Section VI.C.

VI.   Catalysis Research

A. REASONS FOR DOING CATALYST RESEARCH

There are numerous reasons for industrial support of catalysis research. They include the following objectives: basic understanding of chemistry, creation of new catalysts, competive advantage (market shares), patent position, solving plant problems, and improving profits.

An understanding of the chemistry and engineering details of a chemical process is important to a company that uses or plans to utilize the process. Good science builds the reputation of the company and its scientists. This reputation is valuable in recruiting and in customer relations. A standard of excellence in research is desirable for the morale of the scientists and their personal development. A measure of the importance associated with basic catalysis research is the number of scientific papers published by industrial scientists and the excellent support given by research directors to publications such as this work.

Catalyst development can be merely a search for a catalyst, but a higher objective has always been to create a catalyst based on scientific principles. Progress is being made toward this objective and is fostered by the basic catalysis research funded by corporations.

A new or improved catalyst is often the basis of a competitive manufacturing cost advantage. Contributions to a lower manufacturing cost can come from any of the following: (1) reduced equipment costs, (2) reduced feedstock costs, (3) lowered utility costs, (4) improved stream factor, (5) increased by-product credits or reduced by-product debits, and (6) decreased catalyst usage.

Catalyst costs themselves are usually an insignificant portion of manufacturing costs, typically ranging from 0.1¢ to several cents per pound for commodity chemicals.

The competitive advantage may also take the form of a superior product because of the purity, isomer distribution, etc., of the final product. In the catalyst preparation industry itself there is intense competition to develop superior catalysts. Because of the leverage created by a superior catalyst in regard to manufacturing costs, a proven high-performance catalyst often captures a sizable market share. If it is not continually improved by research, however, it will be challenged by a superior competitive product and rapidly lose its market share. Thus, research in catalysis is initiated to preserve or capture a particular market share in the catalyst and chemicals business area.

Technical information is a valuable asset to a company and is developed at considerable expense. It can be kept as a trade secret or patented. A patent gives the owner the right to exclude others from making, using, or selling the invention for a period of time — 17 yr in the United States. Patents serve not only to protect operations but may be licensed or sold to other companies for significant income. A patent may cover a process or method, a product or composition, or an apparatus. Because of their novel composition or method of preparation catalysts are often patented. In addition to being novel, an invention also must have utility and not be obvious. Because catalysis is often an art as well as a science, patents based on catalysis are not obvious and utility is easy to demonstrate. Composition-of-matter patents are the most valuable. An excellent example of a patent excluding others from practice is Mobil Oil Corporation's composition-of-matter patent on ZSM-5 shape-selective zeolite [12]. The patent describes "a crystalline aluminosilicate zeolite having a composition in terms of mole ratios of oxides as follows:

$$0.9 \pm 0.2 M_{2/n}O : Al_2O_3 : YSiO_2 : ZH_2O,$$

wherein M is at least one cation having a valence n, Y is at least 5 and Z is between 0 and 40, said aluminosilicate having the x-ray diffraction lines of Table I of the specification." Such a patent places the owner in a very favorable licensing position.

Patents, like publications in academia, are a status symbol in industrial research for both the individual and the company. Catalyst-related patents are a sizable fraction of the chemistry patents in the *Central Patents Index* of Derwent Publications, Ltd., and the *U.S. Patent Gazette.*

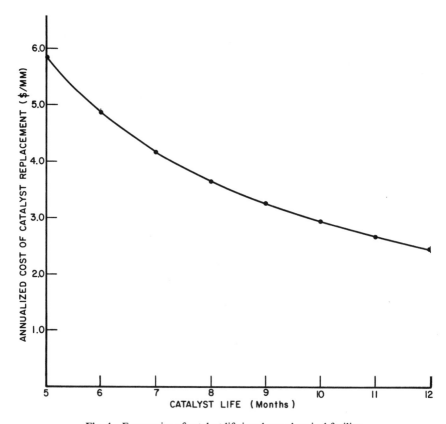

Fig. 4. Economics of catalyst life in a large chemical facility.

Catalysis research may be defensive in nature. The objective may be to ensure continued operation of a commercial facility. This may be done by catalyst and feedstock evaluation to solve plant operational problems before or as they arise. The incentive is great to solve these problems rapidly. Downtime at a large reactor facility is worth 2 – 3 man-years of research time per day. Catalyst research is needed to give the scientist the background data required to diagnose plant operational problems rapidly. The risk of not doing catalyst research is simply too high for a plant manager to accept on a long-term basis.

Another item leading to increased profitability is improved catalyst life. In Fig. 4 an example is given of a 1-billion-lb/yr chemicals plant where catalyst changes require a 2-week shutdown. A study of the annualized costs indicates there is a variable return depending on the current state of the art. There is not much incentive to develop a new catalyst for the application chosen for comparison if the present catalyst has a 9-month life and the new catalyst has a 12-month lifetime. However, if the state-of-the-art catalyst has a 4- to 6-month life range, the incentive for catalyst research is great. The other factors important to the particular analysis include (1) downtime for catalyst change, (2) size of the commercial unit, (3) catalyst replacement costs, (4) the value of lost production capacity, (5) market conditions, and (6) an estimate of the difficulty of preparing a superior catalyst.

B. HINDRANCES TO CATALYST RESEARCH

Catalysis is a complicated art and science. Although the benefits of catalysis research have been enumerated, there are a number of hindrances to industrial catalysis research. Each corporation has evolved a policy consistent with its objectives and limitations.

The study of catalysis requires a sizable technology base, both in scientific expertise and instrumentation techniques. There is a wide spectrum of involvement depending primarily on a corporation's decision whether to develop its own catalyst technology or to license this technology from others. The evaluation of technology requires less labor, and licenses are generally available for commodity chemicals synthesis at reasonable rates. New products require a greater effort, in the area of catalyst development. Corporations with a reputation for developing new catalysts and products usually have a larger research staff and a greater variety of sophisticated spectroscopic instrumentation techniques available as resources than corporations that license technology. However, size alone does not ensure innovation, and even small research organizations can develop specific catalysts.

The two extremes in catalysis research objectives have been presented. There is a continuum of alternatives between theoretical work on new catalysts and the decision to evaluate only commercially available catalysts. Catalyst manufacturing companies often work with an inventor to commercialize the catalyst recipe. Rental of spectroscopic instrument facilities is possible when the workload does not justify the capital expenditure. The wide range of expertise needed to create a catalyst is a hindrance to catalyst research, but selected contract research can minimize the limitations of resources.

C. CATALYST MANUFACTURING

The develop-or-buy consideration regarding catalysts depends on whether the inventor has a large volume demand and facilities for catalyst manufacturing. Some oil and chemical companies have gone into the catalyst preparation business. Others who determined they could not economically produce catalysts themselves have retained the services of an outside manufacturer who makes proprietary catalysts. This is developed in depth in Chapter 2. A list of companies who actively sell catalysts in the United States is given in Table V.

Catalyst companies specialize in particular catalysts for which they have technical expertise and a historical market position. Catalyst companies deal with factors that have no counterpart in laboratory preparations: scale-up of operations — precipitation, mixing, filtration, drying, forming, and calcination; continuous unit operations; energy conservation and environmental control; and optimum use of production facilities.

D. METALS SUPPLY AND COST

Catalyst selection and development must include an evaluation of metals supply and cost. This is particularly true of precious and strategic minerals. As examples the composition of reforming catalysts is adjusted to provide the most cost-effective catalyst composition, and nickel rather than the more expensive cobalt is now the choice of refineries in the hydrodesulfurization of crude oil.

Major sources of strategic materials are shown in Fig. 5. The materials most critical to the catalyst industry that are on the strategic list are chromium, cobalt, manganese, and platinum group metals.

In the design and selection of a commercial catalyst one must consider the volume of metal to be used in relationship to the supply and the natural mix

TABLE V

Major Catalyst Companies in the United States

Activated Metals	Haldor Topsoe, Inc.
Air Products and Chemicals	Harshaw Chemical
American Cyanamid	Katalco
Armak Chemical Division	Matthey Bishop, Inc.
BASF Wyandotte	Montedison, USA, Inc.
Calsicat Division, Mallinckrodt	Nalco Chemical
Dart Industries	Oxy-Catalyst, Inc.
Davison Chemical Division, W. R. Grace	Shell Chemical
Degussa Corporation	Stauffer Chemical
Englehard Minerals and Chemicals Corporation	United Catalysts, Inc.
Filtrol	Universal Oil Products,
Halcon Catalyst Industries	Inc.

Fig. 5. Sources of strategic materials. Adapted with permission from C & EN, May 11, 1981, p. 21. Copyright 1981 American Chemical Society 1981.

of the metals mined. For example, the normal mix of platinum to rhodium is 19:1. A major problem in three-way catalytic converters for automobile exhaust treatment has been that the optimum ratio of platinum to rhodium for this application is closer to 5:1. Major use of such a mixture would have resulted in an overproduction of platinum and an intense shortage of rhodium. A solution has been to consider the converter as having two parts: one section with a high Pt:Rh ratio and the other with a lower Pt:Rh ratio. Overall ratios more in balance with the natural mix are thus obtained.

Precious metals are recycled, but most base metal catalysts are not, including some that are quite high in nickel, cobalt, copper, and chromium. The first problem in recycling is the presence of residual organics on the spent catalyst.

The removal of organics adds another cost to the recovery process but is generally required because they interfere in aqueous metal ion separation schemes and are a water pollution problem. The second factor limiting metal recycle is that many catalysts contain mixtures of metals not found coexisting in the natural ores routinely processed to yield a particular metal. An example is copper chromite catalyst.

A third factor is volume and the need for a transportation network to collect catalysts for metals recovery. The spent catalyst's geographical dispersion is a primary factor in potential metals recovery from automobile exhaust catalysts. The quantity of precious metal per automobile is about 0.05 troy oz. Recovery from alumina supports would be practical if there existed a collection and transportation system that could deliver spent catalyst containers to a central location.

The final reason most metals used in catalysts are not recycled is economic. Metal prices in general have not kept pace with oil and general inflation price increases. New relatively low-volume processing plants cannot compete with large existing metal refining facilities under current market conditions. Factors such as stability of supply, dependence on other countries, balance of payments, and strategic metals considerations, should increase metals recycling in the future.

VII.   Effects of Catalysis on Life-Style

A.   PLASTICS

One of the largest changes in our lives this century has been the introduction of large-volume plastics into the consumer market. The production of most polymers involves catalysts either in the polymerization itself or in the monomer synthesis. Some of the largest volume plastics (Table VI) and the types of catalysts (Table VII) employed in their synthesis are given as examples. In the case of polyethylene and polypropylene there are hundreds of Ziegler–Natta catalyst modifications in the patent literature.

B. TRANSPORTATION

Modern society is highly mobile and depends on the rapid transit of people and commodities. Transportation vehicles incorporate many of the plastics described in the previous section. These materials have been substituted for metal to reduce weight and cost. Another major change has occurred in tires. Catalysts have allowed the preparation of synthetic rubber and fibers that add strength to tires.

Better fuels for the transportation industry have been made available through refinery catalyst developments. Some refinery processes are described in detail in subsequent chapters. Without catalysts to convert crude oil into high-octane fuels efficiently, our transportation system would be severely limited and our life-style significantly impacted.

C. DETERGENTS

Common household detergents that are biodegradable and effectively clean our clothing and dishes are an example of the subtle involvement of catalysis in our life-style. The first synthetic detergents were produced in Germany during World War I when animal fats were not available for soap manufacture. A wide variety of surfactants are now produced. Our focus is on the catalysts used to prepare the building blocks for surfactants and the actual synthesis of the active ingredients. Only large-volume surfactants are described in Figs. 6–8. Alkyl sulfonates and olefin sulfonates are made with sulfur trioxide, which provides adequate acid catalysis.

D. FOOD SUPPLY

Catalysis has played an important role in increasing crop yields to meet the food demands of an increasing world population. Fertilizers, pesticides, and herbicides have been used to increase yields of agricultural commodities. Chemicals have made it possible to grow more on less land with less input of labor and energy.

TABLE VI

Some Important Industrial Polymers

Name of polymer	Monomer	Polymer
Polyethylene	$CH_2=CH_2$	$\cdots CH_2-CH_2 \cdots$
Polypropylene	$CH_3-CH=CH_2$	$\cdots \overset{\displaystyle H}{\underset{\displaystyle CH_3}{\overset{\vert}{\underset{\vert}{C}}}}-CH_2 \cdots$
Polyvinyl chloride (PVC)	$CH_2=CHCl$	$\cdots \underset{\displaystyle Cl}{\overset{\vert}{CH}}-CH_2 \cdots$
Polytetrafluoroethylene Teflon, Halcon, Fluon, Hostaflon, Algoflon, Polyflon, Soreflon, Fluoroplast	$CF_2=CH_2$	$\cdots CF_2-CF_2 \cdots$
Polyacrylonitrile Acrilan, Creslan, Orlon, Zefran, Dralon	$CH_2=CHCN$	$\cdots \underset{\displaystyle CN}{\overset{\vert}{CH}}-CH_2 \cdots$
Polyacrylic acid	$CH_2-CH-COH$ with $\overset{\displaystyle O}{\|}$	$\cdots CH_2-\overset{\displaystyle HO-C=O}{\underset{\displaystyle H}{\overset{\vert}{C}}}- \cdots$
Polymethylmethacrylate	$CH_2=\overset{\displaystyle CH_3}{\overset{\vert}{C}}--\overset{\displaystyle O}{\overset{\|}{C}}-OCH_3$	$\cdots \underset{\displaystyle \underset{O}{\overset{\|}{C}-OCH_3}}{\overset{\displaystyle CH_3}{\overset{\vert}{\underset{\vert}{C}}}}-CH_2 \cdots$

Polymer	Monomer	Repeat unit
Polystyrene	$CH_2{=}\underset{H}{\overset{C_6H_5}{C}}$	$\cdots{-}\underset{H}{\overset{C_6H_5}{C}}{-}CH_2{-}\cdots$
Polyisoprene	$CH_2{=}\overset{CH_3}{C}{-}CH{=}CH_2$	$\cdots{-}CH_2{\overset{CH_3}{C}}{=}CH{-}CH_2{-}\cdots$
Polybutadiene Cis-4, Budene, Diene, Cisdene, Ameripol CB	$CH_2{=}CH{-}CH{=}CH_2$	$\cdots{-}CH_2{-}CH{=}CH{-}CH_2{-}\cdots$
Polyformaldehye Ultraform, Celcon, Delrin, Hostaform, Tenoc, Duracon, Kenmetal	$\underset{HCH}{\overset{O}{\|}}$	$\cdots{-}\underset{H}{\overset{H}{C}}{-}O{-}\cdots$
Poly-1-butene	$CH_2{=}\overset{C_2H_5}{CH}$	$\cdots{-}CH_2{-}\overset{C_2H_5}{CH}{-}\cdots$
Polyphenylene oxide PPO, Noryl Polyamide	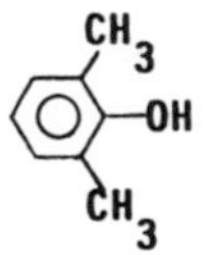 2,6-dimethylphenol (CH₃, OH, CH₃ ring)	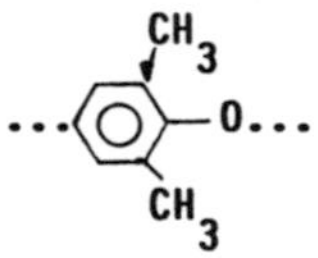 aromatic ring with CH₃, O, CH₃
Nylon For example, Nylon 66	$HOOC(CH_2)_4COOH + H_2N(CH_2)_6NH_2$ adipic acid hexamethylene diamine	$\cdots\underset{O}{\overset{O}{\|\|}}C(CH_2)_4\overset{O}{\overset{\|\|}{C}}{-}\overset{H}{N}(CH_2)_6{-}\overset{H}{N}{-}\cdots$

Table VI (continues)

23

TABLE VI (*Continued*)

Name of polymer	Monomer	Polymer
Poly(ethylene terephthalate) (Dacron)	$CH_3OOC(C_6H_4)COOCH_3 + HOCH_2CH_2OH$	$\cdots C-(C_6H_4)-COCH_2CH_2O\cdots$ (with $\overset{\text{O}}{\|}$ and $\overset{\text{O}}{\|}$, $O-CH_3$, CH_3)
Polycarbonate Lexan, Merlon, Makrolon, Makrofol, Panlite, Jupilon, Touflon	Bisphenol A + phosgene	structure with two C_6H_4 rings, central $C(CH_3)(CH_3)$, $O\cdots$ and $\cdots OC$ with $\overset{\text{O}}{\|}$

TABLE VII

Catalyst Types in Polymer Synthesis

Polymer	Major uses	Catalyst type
Polyethylene	Film and sheet, injection molding, blow molding	Chromia (Phillips), titanium trichloride (Ziegler–Natta)
Polypropylene	Film and sheet, injection molding, blow molding	Titanium (Ziegler–Natta) (isotactic)
Polyvinyl chloride	Molding and extrusion, sheeting, flooring, wire coating, adhesives and paints, film, furnature, clothing	Peroxide initiator; monomer preparation requires $CuCl_2$–KCl–Al_2O_3 oxychlorination catalyst
Polytetrafluoroethylene (Teflon)	Fluid handling, packings, electrical wire coating, pipe and hose, nonstick surfaces	Na, K, NH_4 peroxydisulfate polymerization agent
Polyacrylonitrile	Fibers, fabrics, elastomers, plastics	Monomer preparation: $C_3H_6 + NH_3 + \frac{3}{2}O_2 \rightarrow CH_2{=}CHCN + 3H_2O$, bismuth phosphomolybdate catalyst
Polyacrylics	Copolymer with butadiene, fibers, plastics	Monomer synthesis by oxidation of propylene, BiO_3/MoO_3; $CH_2{=}CH_2 + 2CO + O_2$, $PdCl_2/CuCl_2$
Polymethylmethacrylate	Plastics, sheets, paints, textiles	Monomer prepared using acetone, methanol, and hydrogen cyanide $$2H_2 + CO \xrightarrow{Cu \text{ or } Zn-CrO_2} CH_3OH$$ $$CH_4 + NH_3 \overset{Pt}{\rightleftharpoons} HCN + 3H_2$$
Polystyrene	Insulation, packaging, injection molding, extrusion, meat trays, film	Free radical initiation polymerization Isotactic polystyrene made with Ziegler–Natta catalyst. Monomer made by dehydrogenation of ethyl benzene over Fe_2O_3–KOH–Cr_2O_3
Polyisoprene	Synthetic rubber blending with natural rubber or cis-1,4-polybutadiene	β-$TiCl_3$, Ziegler–Natta, or alkyllithium
Polybutadiene	Rubber, especially copolymer with styrene (SBR) and acrylonitrile (NBR)	Ziegler–Natta catalyst for polymerization. Dehydrogenation of butane or butene

Table VII (continues)

TABLE VII (*Continued*)

Polymer	Major uses	Catalyst type
Polyformaldehyde	Engineering plastic	BF_3 catalyst used in polymerization. Monomer made by air oxidation of methanol over $Fe(MoO_4)_3$ or silver gauze
Poly-1-butene	Pipe, film	Ziegler–Natta catalyst (isotactic)
Polyphenylene oxide	Engineering plastics	Cu^+, amine polymerization catalyst—also Mn^{2+}
	Injection molding, pipe and rod, film, sheet, slab	Magnesium oxide catalyst to prepare 2,6-xylenol from methanol and phenol
Nylon	Fiber, fabric, carpets, yarns	Melt spinning
Poly(ethylene terephthalate) (Dacron)	Fiber, fabric, carpets, yarns	Antimony catalyst (many others also possible), dimethyl-terephthalate
	Adipic acid synthesis:	
	$Cyclohexane + 5O_2 \xrightarrow{\text{cobalt acetate}} 2 \text{ adipic acid} + 2H_2O$	
	Terephthalic acid synthesis:	
	$p\text{-Xylene} \xrightarrow[Br_2 \text{ catalyst}]{O_2} \text{terephthalic acid}$	
	Ethylene glycol synthesis:	
	$CH_2=CH_2 \xrightarrow[Ag]{\frac{1}{2}O_2} CH_2 \overset{O}{\overbrace{\quad}} CH_2 + H_2O \xrightarrow{H^+} HOCH_2CH_2OH$	
	$2CH_2=CH_2 + 2H_2O + O_2 \xrightarrow[\text{or manganese acetate + potassium iodide catalyst}]{TeO_2 \text{ promoted by Br compounds}} HOCH_2CH_2OH$	
Polycarbonates	Engineering thermoplastics, extrusion, film, blow molding	$AlCl_3$ polymerization catalyst

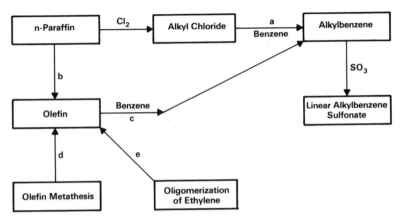

Fig. 6.   Linear alkylbenzene sulfonates. a, AlCl$_3$ catalyst; b, Pt/AlO$_3$ catalyst; b, Pt/Al$_2$O$_3$ catalyst; c, HF catalyst; d, CoO–MoO$_3$–Al$_2$O$_3$; e, aluminum alkyls with heat- or nickel-catalyzed alkyl displacement.

About 80% of the ammonia produced worldwide is used as fertilizer. Ammonia is formed from the catalytic reaction of nitrogen from air and hydrogen from natural gas.

The synthesis of herbicides and pesticides also involves catalysts. The markets are highly fragmented, but the agrichemicals research area is one of high activity.

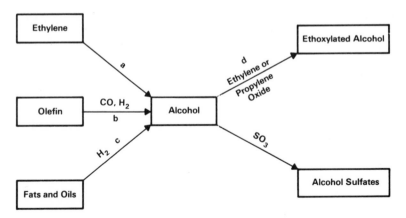

Fig. 7.   Alcohol-derived surfactants. a, Trialkyl aluminum; b, cobalt or rhodium oxo catalyst; c, copper chromite catalyst; d, NaOH, Ba(OH)$_2$, or Sr(OH)$_2$.

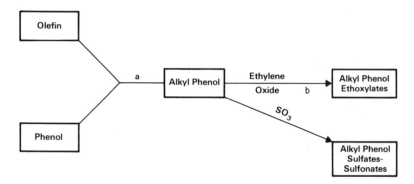

Fig. 8. Alkyl Phenol Derived Surfactants. a, HF or strongly acidic ion-exchange resin; b, NaOH catalyst.

E. ENERGY CONVERSION AND CONSERVATION OF RESOURCES

In the future we may be faced more with a shortage of energy in a particular form than with a total shortage of energy sources. Catalysts will play an increasingly important role in the synthesis of chemicals and fuels from coal, shale oil, and tar sands.

Conversion of energy from one hydrocarbon form to another basically involves hydrogenation–dehydrogenation reactions. Coal and shale oil are deficient in hydrogen and are liquefied by catalytic hydrogenation.

Alternatively the conversion can be accomplished by gasification to H_2 and CO followed by a synthesis reaction to alcohols or hydrocarbons after an appropriate water gas shift reaction [Eq. (2)] to form the required amount of hydrogen.

$$CO + H_2O \xrightleftharpoons{\text{Catalyst}} CO_2 + H_2 \tag{2}$$

Energy conservation will become more important. Improved catalysts for commercial processes have the potential for reducing operational pressures and temperatures, thereby saving energy.

E. ENVIRONMENTAL EFFECTS OF CATALYSTS

Catalysts have found extensive use in pollution control. The catalysts used to reduce automotive emissions have helped control the hydrocarbons, carbon monoxide, and nitrogen oxide levels in the atmosphere. The same type of precious metal-based catalysts are also useful in reducing hydrocarbon emissions from vent streams in chemical operations.

While catalysts can help clean up the environment, at the same time they can themselves pose environmental problems. Solid waste disposal is a major problem in the 1980s. It has received publicity because of mistakes made in the handling of hazardous wastes in the past. The question of what constitutes safe disposal is still being debated. New regulations covering shipping, packing and storing wastes have already been set in the United States, with stiff legal penalties for failure to comply.

The disposal costs and the recycle value of metal catalyst components combine to give catalyst regeneration a promising future. Catalysts containing more toxic heavy metals that may be extracted into groundwater (e.g., chromium) will stimulate research for alternative catalyst systems that do not require expensive or hazardous waste disposal.

Catalyst preparation itself often involves aqueous salt solutions from precipitation reactions and metal ion-containing waste streams. Ion exchange has been very helpful in solving metal contamination problems.

In calcination, the decomposition of nitrates, sulfates, chlorides, etc., produces vapors that must be scrubbed to ensure air quality control. The acidic aqueous solutions are neutralized and the salts concentrated in holding ponds.

Ion-exchange resins have been developed to reduce precious metals selectively. Amborane reductive resins prepared by Rohm and Haas Company can reduce precious metals such as Au^{3+}, Pt^{2+}, Pt^{4+}, Rh^{3+}, and Ir^{3+}, retaining the reduced metal within the water-insoluble polymeric resin. The metals can then be recovered by slowly roasting the metal-containing beads of polymer. The capacity for reduced metal is $1-2$ g metal per gram of dry resin. The stoichiometry of the reduction is given in Eq. (3).

$$n - \textcircled{P} - BH_3 + 6\ M^{n+} \longrightarrow \textcircled{P} - H^+ + 6\ M^\circ + 5n\ H^+ + nB(OH)_3 \qquad (3)$$

Amborane Resin	Metal Ion	Resin Loaded With Reduced Metal	Boric Acid

Methods of metal recovery will continue to be developed given the emphasis on water quality and the impact of metals cost and availability.

References

1. E. J. Rosinski, T. R. Stein, and R. H. Fischer, U.S. Patent 4,082,695 (1978).
2. R. D. Christman, G. E. Elliott, and G. Guelfi, U.S. Patent 3,730,879 (1973).
3. P. Sabatier, "Catalysis in Organic Chemistry." Van Nostrand, New York (1922).

4. A. Mittasch, In "Advances in Catalysis" (Vol. 2), pp. 81 – 103. Academic Press, New York (1950).
5. B. Luberoff, *Chem Tech.* page 8, (1981).
6. K. Ziegler, *Brennst. Chem.,* **35,** 321 (1954); Belgian Patent 527,736 (1954).
7. J. Boor, Jr. "Ziegler – Natta catalysts and polymerizations," pp. 93 – 100. Academic Press, New York (1979).
8. H. Coover Jr., F. Joyner, and N. Shearer Jr., Belgian Patent 577,216; also see U.S. Patent 3,549,608, issued to H. W. Coover and F. B. Joyner (1970).
9. J. Stevens, and M. George, (1976). Belgian Patent 757,847; Solvay & Cie (1976).
10. K. Tsubaki, H. Morinaga, Y. Matsuo, and T. Iwabuchu, UK Patent Application GB2,020,672A (1979).
11. B. Leach, *J. Org. Chem.* **43,** 1794 (1978).
12. R. Arganer, and G. Landolt, U.S. Patent 3,702,886 (1972).

CHAPTER 2

Catalyst Scale-up—Pitfall or Payoff?

EUGENE F. SANDERS   EDWARD J. SCHLOSSMACHER*

Mallinckrodt, Inc.
Calsicat Division
Erie, Pennsylvania

I. Introduction and Definitions

The art, technology, and business of catalyst scale-up are widely misunderstood by the majority of laboratory practitioners. Understanding what is involved and thinking in terms of scale-up even in the laboratory greatly facilitate the process and may even mean the difference between success and failure.

It is important to define what we mean by catalyst scale-up and other terms we shall use in the following sections:

(1) Catalyst scale-up—a process whereby a catalyst previously made in the laboratory, say in gram quantities, is to be manufactured in 100- or 1000-lb quantities at reasonable rates and economically.

(2) Prototype catalyst—a catalyst prepared using commercial preparation facilities.

(3) Pilot plant—a small-scale catalyst preparation facility, characterized by small-volume equipment and high labor, technical consultation, and analytical intensity, that simulates the "rate" behavior and other key operations of a commercial catalyst preparation facility.

(4) Commercial scale—manufacturing characterized by large-volume

* Present address: Research and Development Department, Ashland Petroleum Company, Ashland, KY, 41101.

equipment, low labor and technical supervision intensity, and intermittent analysis.

(5) Product specification—a list of measurable properties of a given catalyst that are indicative of its required performance and process control during manufacturing.

(6) Surrogate activity test—an economical, stable, and predictable test that is correlatable with the desired activity of a catalyst in commercial operation.

(7) Custom catalyst manufacturer—a firm that has established expertise in catalyst scale-up and whose future financial viability depends on successful commercial-scale production of catalysts of its own invention or of those invented by others.

II. Role of the Custom Catalyst Manufacturer

When faced with the choice of scaling up a newly developed catalyst, either in house or with a custom catalyst manufacturer, the economics, more often than not, make the latter the preferred choice. Some of the reasons why this is true include in-place facilities, both productive and analytical, an established system for scale-up that saves times and increases the chance of success, experienced management, and proprietary know-how that can keep scale-up costs and commercial production costs to a minimum.

This latter point should be a factor in choosing which custom manufacturer to work with. The inventor should try to ensure that the manufacturer has expertise in the areas of interest. Certain firms have established a "center of excellence" in the chemistry and handling of certain catalyst metals, e.g., palladium, platinum, nickel, and vanadium, or certain substrates, e.g., alumina, silica, zeolites, and carbon. Further, a firm may have particular expertise in specific catalyst manufacturing unit operations.

The range of unit operations utilized in catalyst manufacture is quite broad and includes the following: mixing, impregnation, precipitation, drying, filtration, fluid and fixed bed reduction, calcining, extrusion, pilling, and spheroidizing. The efficient management and adaptability of existing facilities can mean substantial product cost savings.

III. The Scale-up Process

The following description of the scale-up process assumes that a custom manufacturer has been brought into the picture but, apart from the legal and commercial negotiation aspects, the process described is applicable to an in-house scale-up also. Generally speaking, the sooner a custom manufacturer is brought into the effort the better, because he can prevent an occurrence of the pitfalls discussed later.

The first step after reaching a general agreement to work together is to arrange a unilateral or bilateral confidentiality agreement (Table I). This serves as a basis for disclosing information and data required for and during the scale-up process and to protect the proprietary information of both parties. Any reputable firm should be willing to enter into such an agreement, and there should be no reluctance on the part of the inventors to reveal all pertinent information. Many custom manufacturers go to great pains to keep these confidences, because their future viability depends on a sound reputation.

The next step is to make a judgment about the chances of a successful scale-up. One or more duplications of the laboratory preparation will be required as the custom manufacturer attempts to understand the chemistry involved.

It is absolutely vital that the question, How will success be measured? be answered. The inventor needs to have a clear understanding of what constitutes the catalyst he is seeking. At a bare minimum this includes physical properties, such as size, strength, and density, and the chemical composition of both major components and impurities. Ideally, there are measurable physical and chemical characteristics that relate to catalytic behavior; however, in actual practice, such relationships are often obscure. It is recommended that some measure of activity using either the actual conversion process or a surrogate activity test be agreed upon (see Fig. 1). These specifications must describe the catalyst requirements yet must be reasonable. The payoff of a successful scale-up program is a timely, economical, and reproducible catalyst that meets the user's needs. It will not be the ultimate catalyst prepared by the ideal process.

The next step should be the formulation of a conceptual preliminary

TABLE I

Elements of a Common Confidentiality Agreement

Names of the parties to the agreement
Definition of what constitutes the information to be exchanged
Commitment to keep information confidential and limit its use for other purposes
Time period (term)
Exceptions to commitment
   Where information is already known
   Where information becomes publicly known
   Third-party disclosure
   Where information is independently developed

Fig. 1. Typical fixed bed reactor test unit for catalyst evaluation with a remotely located control panel.

commercial process and based on this, an estimated production cost/price estimate. The sooner the issue of catalyst price is understood the better it is for all parties. The inventor should also seek at this time a commitment from the custom manufacturer as to the cost of the scale-up program, since many vendors expect these costs to be borne by the inventor even if commercial-scale manufacture is never realized.

Fig. 2. Typical glass-lined pilot plant system for precipitating catalyst formulations in a corrosive environment. A fiberglass fume scrubber appears in the background.

If the program and production cost estimates so dictate, the next step will be simulation of the catalyst preparation in the laboratory and then at a pilot plant. It is important that all the key steps be simulated and that perturbations in process variables and raw material quality be investigated. An ideal pilot plant is one that has been itself scaled down from an existing or proposed commercial-scale facility (see Fig. 2).

Several pilot plant preparations may be made, but eventually a commercial-scale process is decided upon and an updating or revision of the catalyst manufacturing cost made. With these established, the inventor and manufacturer are then ready to negotiate commercial arrangements and a timetable. These arrangements run the gamut, such as take or pay, fixed profit, cost plus, toll manufacture with capital underwriting, and supply contract. Other issues that should be addressed at this time are shown in Table II.

The day finally arrives when a prototype production run is made. After complete testing and verification of specifications, commercial production begins and another catalyst has been successfully scaled up.

There are pitfalls along the way that can lead to a costly or even an unsuccessful scale-up. Let us consider some of these problems and how to avoid them.

IV.   Pitfalls To Be Avoided

Many pitfalls can be traced back to the laboratory preparation, invention phase of the process. In many ways the laboratory is an ideal environment. There are fewer environment-related problems. Mixing is done on a small scale, chemicals are pure, and overnight processing is usually no problem. The effective use or recycle of raw materials never seems to be an issue. Let us examine some of these and other issues.

The raw materials or substrates used in catalyst preparation are quite often specialty products themselves. The suitability of commercial grade raw materials in the preparation and the effect of the impurities in such material should be checked. If at all possible, one should avoid becoming involved with only one supplier. If this is not possible, a confidential

TABLE II

Elements of a Common Catalyst Supply Agreement

Quantity
Price and terms
Escalation arrangements
Production, lead time, and delivery commitments
Specifications, including sampling arrangements and analytical methods
Packaging requirements
Special storage and shipping arrangements
Amortization of specialized equipment
Force majeure clause

arrangement should be made with him and his sourcing capabilities and pricing policies discussed.

The initial processing step is most commonly either an impregnation or precipitation step. Impregnation of a preformed carrier with the active catalyst species is generally preferred for low levels, less than 10%, of the active ingredients. In this way, minimum quantities of the ingredients can be well dispersed uniformly throughout the porous carrier or selectively deposited on or near the carrier surface. Identification of the important support properties and the appropriate metal salts and their effect on metal–support interactions, as well as subsequent activation steps, should be made as early as possible in the investigation. A custom catalyst manufacturer can offer good advice on commercial support selection and many have proprietary impregnation technologies.

Precipitation of the catalyst ingredients in the presence of powdered support material is preferred for levels above 10%. The laboratory preparation of precipitated catalysts is especially prone to idealized kinetic behavior. Unfortunately the precipitation step quite often is the most critical one in achieving desired catalyst properties. Experiments in the laboratory should aim at introducing rate-limiting steps that are typical of commercial batch operations. This might mean limiting the mixing power input, rate of reagent introduction, heating or cooling rate, etc.

Precipitation conditions affect not only the catalyst performance but quite often also the ease with which the catalyst is filtered from the reaction medium. Formation of a slime or a gel is a real danger signal that scale-up will be difficult. The chosen precipitation conditions are usually a trade-off that achieves a balance between catalytic behavior, filtering, and handling properties and important physical properties such as attrition resistance for slurry or fluid bed catalysts or forming characteristics for fixed bed catalyst.

A precipitated or impregnated catalyst is generally activated by drying and calcining. Even in the drying or calcining operation the question of rate must not be overlooked. Complex solid state reactions often take place. These reactions have a unique thermal profile or history requirements that are not always possible to duplicate easily with commercial-scale equipment, particularly when laboratory preparations are dried or calcined "overnight" in a small static bed. Thermal gravimetric analysis and differential scanning calorimetry investigations of these steps can yield important scale-up data. A successful scale-up also depends on evaluating the effects of such parameters as agitated drying–calcining and atmosphere control.

Precipitated catalysts are often formed by pilling, extruding, spheroidizing, etc. Care must be exercised that the "demands" concerning density, crush strength, size, porosity, etc., are reasonable. For example, the commercial pilling rate, tooling costs, etc., are adversely affected by an unreasonable specification of such physical parameters.

In addition, there are often interactions between the catalyst-forming feed preparation and the forming operation itself. A successful scale-up may depend on the incorporation of binding materials, pore builders, or pre-forming steps not envisioned at the laboratory stage.

Catalyst activation is often accomplished by reducing the active species to the metal employing a variety of techniques ranging from "wet" reduction using a chemical reagent to "dry" reduction with hydrogen. Each has its own advantages and drawbacks. Here again, many custom catalyst manufacturers have proprietary techniques that already incorporate steps for avoiding the more common pitfalls. Improper activation of experimental catalyst preparations can result in the discarding of a fundamentally excellent catalyst formulation. Other pitfalls can be avoided if the inventor considers such issues as catalyst shelf life requirements and the capabilities of the plant for depassivating pyrophoric catalysts or even *in situ* reduction.

Requirements placed upon commercial manufacturers by Occupational Safety and Health Act (OSHA), Environmental Protection Agency (EPA), and Department of Transportation (DOT) regulations should not be over-looked, even in the laboratory stage. The use of carcinogenic materials, flammable solvents, and toxic chemicals in a catalyst preparation will certainly affect the cost and perhaps the timetable. Premanufacturer notification, hazardous waste disposal, etc., all can create delays in the scale-up effort. These same concerns should be considered, of course, from the perspective of the end use. The disposal of spent catalysts is becoming increasingly difficult and costly. Ways to prevent proprietary aspects of catalysts from being divulged during disposal have to be built into the system.

V. How Can Payoff Chances Be Increased?

The completion of a successful prototype catalyst does not signal the end of the program for the inventor or the catalyst manufacturer. The technology developed and the lessons learned must be incorporated into a system for managing and controlling routine catalyst production.

The process must be frozen at this point, and a master process written. At a minimum, the master process should include the elements shown in Table III. A custom catalyst rarely requires year-round production, nor does it warrant the level of instrumentation and equipment specificity found in today's major chemical processes. The master process provides the continuity and the basis for safe manufacture of a high-quality, reproducible catalyst.

The inventor should ensure that the catalyst manufacturer has an ade-

TABLE III

Elements of a Master Process

Process flowsheet
Process chemistry
Material and energy balances
Raw materials description and specifications
Product characteristics and specifications
Equipment description
Detailed operating instructions
Product packaging and labeling requirements
Discussion of key operating variables
Process control scheme
Chemical, health, and safety hazards and procedures
By product and waste stream handling
Corrosion, contamination, and material of construction

quate quality control plan such as that shown in Table IV, and that a mechanism is provided for documentation and feedback of changes and perturbations such as changes in raw material source or specification, changes or modifications in plant equipment, adjustments in key operating conditions, and process upsets.

Finally, a good deal of thought must be given to product specifications. During the course of catalyst development, a variety of sophisticated instrumentation and specialized characterization techniques are often used to define the catalyst; however, in terms of routine characterization of production quantities, measurements and analytical techniques should be selected

TABLE IV

Elements of a Catalyst Quality Control Plan

Raw materials
  Vendor certification procedure
  Raw material specifications
  Incoming material inspection procedure
In-process control
  Control of key process parameters
  Measurement of important process material properties
  Periodic plant audit procedures
  Formal process amendment procedure
  Records management
Product
  Product specifications
  Inspection plan
  Product certification and release procedure
  Packaging and labeling procedure

TABLE V

Catalyst requirement	Measurement
Activity	Laboratory activity test
	Chemical composition
Selectivity	Trace metal analyses
Lifetime	Metal surface area by gas adsorption
Regeneration stability	Catalyst surface area by gas adsorption
Poison resistance	Pore volume by water absorption
Thermal stability	Pore size distribution by mercury porosimetry
	Differential scanning calorimetry
	Thermal gravimetric analysis
Fixed bed Slurry reactor	
Strength Filterability	Dead weight load crush strength
	Bulk compression crush strength
	Abrasion index
Abrasion resistance Settling rate	Particle size distribution by sieve analysis and Coulter counter
Flow properties Attrition resistance	
	Filter leaf test
Shelf life	Tablet dimensions
	Apparent bulk density

that can be performed rapidly and economically with good precision and can be reproduced in other laboratories. The catalyst manufacturer must certify the catalyst quality based on measurements of its properties. An early agreement on analytical methods, laboratory cross-checks, and lot reserve samples is necessary. Table V is a checklist of some of the more important catalyst requirements and suggested standard measurements for ensuring that the production catalyst has the required characteristics and will achieve the expected performance at the user's plant.

Bibliography

Sydney Andrew, *CHEMTECH* **9**, 180–184, March (1979).
Charles N. Satterfield, "Heterogeneous Catalysis in Practice." McGraw–Hill, New York (1980).
Charles N. Satterfield, *CHEMTECH* **11**, 618–624, October (1981).
A. H., Thomas, and C. P. Brundrett, *Chem. Eng. Prog.* **76**, 41–45, June (1980).
David L. Trimm, *CHEMTECH* **9**, 571–577, September (1979).
Sol W. Weller, *AIChE Symp. Ser.* **70**, 143 (1974).
"Catalyst Handbook" Wolfe Scientific Books, London (1970).

CHAPTER 3

Laboratory Reactors for Catalytic Studies

J. M. BERTY*

Berty Reaction Engineers, Ltd.
Erie, Pennsylvania

Notations

a	Transfer area (m^2)
A	Flow cross section (m^2)
C	Concentration (mol/m^3)
c_p	Heat capacity ($kJ/kg\ K$)
D	Diffusivity (m^2/sec)

* Present address: Department of Chemical Engineering, University of Akron, Akron, Ohio, 44325.

d_p	Particle diameter (m)
F	Feed rate (m^3/sec)
G	Mass velocity (kg/m^2 sec)
h	Heat transfer coefficient (W/m^2 sec)
k	First-order rate constant (sec^{-1})
k_c	Mass transfer coefficient (m/sec)
k_t	Thermal conductivity of fluid (kJ/m sec K)
L	Critical length (m)
r	Rate of reaction (mol/m^3 sec)
t	Reaction time (sec)
T	Temperature (K)
W	Catalyst charge (kg or m^3)
u	Superficial linear velocity (m/sec)
ΔH	Heat of reaction (kJ/mol)
θ	Fraction of pore volume
μ	Viscosity (kg/m sec)
ρ	Density of reacting fluid (kg/m^3)
Da_I	Damköhler group I $= (r/C)(F/W) = k/(u/L) = kt$ (chemical reaction rate/bulk mass flow rate)
Da_{II}	Damköhler group II $= (r/C)/(D/L^2)$ (chemical reaction rate/molecular diffusion rate)
Da_{III}	Damköhler group III $= (-\Delta Hr)/(\rho c_p u T/L)$ (heat generation rate/bulk heat transport rate)
Da_{IV}	Damköhler group IV $= (-\Delta Hr)/(k_t T/L^2)$ (heat generation rate/heat conduction rate)
j_D	Colburn factor for mass transfer $= (k_c \rho/G)(\mu/\rho D)^{2/3}$ (mass transfer rate/bulk mass transport rate) (Sc$^{2/3}$)
j_H	Colburn factor for heat transfer $= (h/c_p G)(c_p \mu/k_t)^{2/3}$ (heat transfer rate/bulk heat transport rate) (Pr$^{2/3}$)
Pr	Prandtl number $= c_p \mu/k_t$ (momentum diffusivity/thermal diffusivity)
Sc	Schmidt number $= \mu/\rho D$ (momentum diffusivity/molecular diffusivity)
Re$_p$	Reynolds number $= d_p G/\mu$ (inertial force/viscous force)
ϕ	Thiele modulus $= Da_{II}^{0.5}$ (reaction rate/pore diffusion rate)$^{0.5}$
Φ	Weisz–Prater criterion $= \eta Da_{II}$ (observable reaction rate/pore diffusion rate)
η	Effectiveness factor $= r_{obs}/r$ (observable reaction rate/rate unlimited by diffusion)

Introduction

Developments in experimental and mathematical techniques in the last 10 yr have initiated a new interest in the development of better laboratory reactors for catalytic studies. Besides the many publications on new reactors for general or special tasks, quite a few review articles have been published on the general subject of laboratory reactors for catalytic studies.

Most of the published reviews on reactors and on the testing of catalysts represent a special viewpoint because the author's own field of interest influenced the paper. Bennett *et al.* [1] reviewed gradientless reactors from the point of view of transient studies. Weekman [2] evaluated various reactors for powdered catalysts used mostly in fluid bed processes as contrasted to particulate catalysts. Doraiswamy and Tajbl's [3] review dealt primarily with fixed bed reactors. Difford and Spencer [4] gave a brief review and recommendations on the use of different reactors for various purposes. Jankowski *et al.* [5] described the construction of various gradientless reactors. Cooke [6] reviewed bench-scale reactors and tried to give a definition of the ideal reactor. Finally Berty [7] reviewed the testing of commercial catalysts in recycle reactors.

All the preceding review papers will serve as general references for the indicated point of view, and where the following discussion cannot go into details the reader should consult these articles.

I.   The Purpose of Laboratory Reactors

A.   DEFINITION OF OBJECTIVES

The task of laboratory reactors in applied industrial catalysis is to find the catalyst that can be used in the most economical way in commercial production units. To find the optimal way of using a catalyst in commercial units is an additional duty for laboratory reactors. In essence laboratory reactors are used in the industry to test catalysts and to develop mathematical models or expressions describing the kinetics of the catalytic process. In both cases the prediction of performance in commercial units is desired. Catalyst testing and kinetic studies are not entirely separate problems, and both tasks must be considered whatever the primary purpose.

The various tasks with increasing kinetic involvement in laboratory testing are the following:

(1)   Quality control testing for replacement in existing processes,

(2)   Testing of various catalysts for improving an existing commercial process,

(3)   Testing of various catalysts for a new process under development,

(4)   Developing a kinetic model for finding optimum conditions for existing processes, and

(5)   Developing a kinetic model for the design basis of a new catalytic process.

Testing catalysts in laboratory reactors is expensive and time-consuming. Therefore, it should be done only for catalysts that have passed other complementary chemical, physical, and physicochemical tests.

B. DEFINITION OF PERFORMANCE CRITERIA

The ultimate measure of the value of an industrial catalyst is its profit-making potential. To estimate this potential one has to know (1) the expected production rate to utilize the investment in the plant, (2) the selectivity to utilize the raw materials, (3) the temperature response to estimate energy costs or credits, (4) the expected lifetime, and (5) the cost.

1. Activity

For the estimation of production rate the catalyst's activity has to be known. The measure of catalytic activity can be the conversion achieved in a tubular reactor under standardized conditions for the reaction or it can be the feed rate required for a standard catalyst volume to achieve a standard conversion. In a tubular reactor, even if the observable conditions of temperature and concentration are at standard values, the effective levels can be different depending on the flow and reaction rate. In recycle reactors operating as continuous stirred tank reactors (CSTRs), the directly measurable rate under fixed conditions or the feed rate required to obtain a definite conversion measures the activity. Because changing the feed rate does not change the internal recycle rate, the confusing effect of flow influence is excluded. In addition, the internal flow rate can be very high where flow effects are absent. This is important for industrial catalysts where the highest rates are demanded that still can be controlled on an industrial scale.

In summary, measures of the activity of catalysts can be expressed in various practical ways, among others:

(a)   The conversion of starting material: $(C_0 - C)/C_0 = X$ at fixed feed rate F, catalyst charge W, temperature T, pressure P, and feed composition C_0.

(b)   The yield of the desired material, expressed as conversion of the starting material to the product on the same basis as (a).

(c)   The average production rate of the desired product, sometimes called the space–time yield, that is, the yield in units such as moles or pounds per unit of catalyst in unit time and under certain standard conditions.

(d)   Any of the above expressions can be normalized to the performance of a standard catalyst, resulting in a simple ratio or percentage expression such as activity $= 1.1$, i.e., the tested catalyst produces 10% more than the standard catalyst.

In all the above expressions the quantity of the catalyst charge W can be expressed in different ways. Although theoretically the performance per unit inside area would be the most meaningful, industrial reactors have a limited capacity by volume or weight for catalyst charge and so these units are better reference values.

2.   Selectivity

In addition to activity, catalyst selectivity must be expressed. Of the many possibilities, the most used are:

(1)   The ratio of moles of starting material converted to the desired product to moles converted to by-products is called the selectivity.

(2)   The moles of starting material converted to the desired product divided by the total moles converted is frequently called the efficiency.

Both activity and selectivity can be expressed as point values, i.e., at a very narrow range of conversion or as integral values always counted as averages between zero and the given conversion.

3.   Temperature Response

Temperature functions of a catalyst or catalytic rates are expressed in an empirical Arrhenius-type form by calculating from the rates an apparent or overall energy of activation. Catalysts with the lowest energy of activation are considered the best for two reasons. First, theoretically catalytic action demonstrates itself by lowering the activation barrier and, second, for exothermic reactions the permissible maximum temperature difference between the catalyst and cooling medium is inversely proportional to the activation energy [8, 9]:

$$\Delta T_{max} = (RT^2/E)(C_0/C),$$

for first-order reactions.

4.   Catalyst Life

Catalyst life in commercial units is one of the most difficult factors to estimate in laboratory reactors. The reason for this is that the cause of catalytic activity decline has more than one source and these sources are usually not well known. Catalyst coking, poisoning, recrystallization, pore plugging, and other decay processes may occur in different proportions in a laboratory and in the various parts of a commercial reactor. Catalyst life can be expressed as follows:

(a)   In time units—for cracking catalysts in seconds and for ammonia catalysts in years,

(b)   In length of time between regenerations or length of the total useful life until final discharge, and

(c)   In total pounds of product made during its lifetime. The useful life of a catalyst is much shorter than the total loss of its activity. At some level of low activity replacement with a more active new catalyst becomes economical. This point depends on many other economical factors and changes with market conditions.

II.   Transfer Processes and Rate-Limiting Steps

Industry needs the most active catalysts it can handle effectively. The limits on activity are sometimes set by trade-offs with selectivity demands. In addition to knowing how active and selective a catalyst is, it is important to know the nature of the rate-controlling phenomenon. With highly active catalysts this can be a transfer process that may be flow-dependent and consequently can differ in laboratory and industrial reactors if they operate under different flow regimes.

A.   EXTERNAL LIMITATIONS

There are a few published criteria for estimating the importance of the transfer processes at the outer surface of catalysts [10, 11]. These criteria are usually complicated and necessitate the knowledge of more kinetic information than is generally available. In addition, the values of the dimensionless criteria are not numbers that can be easily appreciated. For example, Carberry's "generalized nonisothermal external effectiveness" equation is [10]:

$$\eta = (1 - Da_{II}/Sh)^n \exp\{-\epsilon[(Nu/Da_{IV}) - 1]\}.$$

To use this equation one has to know the Arrhenius number $\epsilon = E/RT$ and the order of reaction n, and these values may not be known when the effect of transfer processes needs to be evaluated.

From the experimental data and a few physical properties the concentration and temperature differences at the outside surface can be estimated as follows:

$$Re_p = d_p G/\mu,$$

where $G = u\rho$. Then

$$j_D = j_H = 1.15/Re_p^{0.5}$$

[12] from the transport coefficients

$$k_c = j_D u(Sc^{2/3}) \quad \text{and} \quad h = j_H \rho c_p (Pr^{2/3})$$

and finally

$$C_s - C = r/k_c a \quad \text{and} \quad T_s - T = -\Delta H_r/ha.$$

The concentration and temperature differences when normalized to the values inside the transfer film become

$$(C_s - C)/C = Da_{II}/Sh \quad \text{and} \quad (T_s - T)/T = Da_{IV}/Nu.$$

These last expressions, although more general, are less practical for use in laboratory tests. For more on dimensionless numbers see Boucher and Alvez [13, 14].

If the concentration and temperature differences are less than some arbitrary limit such as 1°C or $\frac{1}{20}$ the concentration, these effects can be neglected.

B. INTERNAL GRADIENTS

The Thiele modulus [15] is the square root of the ratio of reaction rate to diffusion rate:

$$\phi = L(k/D_e)^{0.5} \quad \text{and} \quad \phi = Da_{II}^{0.5}.$$

The Thiele modulus is the square root of the second Damköler number evaluated for pore diffusivity. The first-order kinetic constant k is seldom known, and for higher-order rates the estimate becomes complicated; so it is more practical to use the Weisz–Prater [16] criterion which contains only observable variables:

$$\Phi \equiv \eta\phi^2 = (r/C_s)(L^2/D_e).$$

If Φ is much greater than 1, significant diffusional limitation is present. This criterion is strictly valid for the case where no significant internal temperature gradient exists inside the catalyst pellet. Because this is the usual case except for very exothermic reactions in large pellets, this equation can be used reliably.

The case when Φ is much less than 1 means an absence of diffusional limitations, but only for the simplest first-order irreversible kinetics, and does not have general validity. Because many reactions behave in a limited range like first-order reactions, or for the approximation of rate expressions results are force-fitted to first-order models, it should be used with caution. Better and necessarily more complicated criteria require a knowledge of kinetic details that are seldom available and usually questionable. Mears [11] reviewed and discussed several of these criteria, including heat effects. Practical testing in a laboratory reactor for pore diffusion can be done on commercial catalysts of the original size and on particles crushed to smaller sizes. If a diffusional limitation exists, rates should be inversely proportional to particle size. Nondestructive testing can be done by studying the effect of inert gas partial pressure on the rates at constant partial pressure for reactants and products [7, 17].

III. Requirements for a Scale-down to Laboratory Reactors

A. DEFINITION OF THE IDEAL LABORATORY REACTOR

For the study of fast and exothermic reactions, such as oxidations, the requirements were set as follows [18]:

(1) "The catalytic reaction should occur on the catalyst at the same mass velocities as are practiced commercially, assuring the same kinetic regime and heat and mass transfer conditions between catalyst and gas. The mass velocity of the gas should be well defined."

(2) "It should even be possible to exceed the mass velocities practical in commercial operation and run under better heat and mass transfer conditions or at higher chemical rates."

(3) "It should be possible to use the reactor under pressure hermetically closed to allow the use of radioactive tracers under clinically clean conditions and to avoid catalyst poisoning."

(4) "It should use small quantities of both reactants and catalysts."

Discussing laboratory reactors and their limitations, Weekman [2] listed the following critical performance factors (renumbered here for easier reference later on):

(5) Sampling and analysis of product composition,
(6) Isothermality,

(7) Residence contact time measurement,
(8) Selectivity time-averaging disguise,
(9) Construction difficulty and cost.

Weekman was mostly concerned with powdered, rapidly decaying catalysts like those used in catalytic cracking.

The ideal reactor, as defined by Cooke [6], does not satisfy the requirements for well-known mass velocities and transfer coefficients. Cooke adds the following attributes for his ideal reactor:

(10) Well-mixed conditions,
(11) Isothermality, gradientless conditions,
(12) Easy computation of rates,
(13) High conversions and ease of model discrimination,
(14) Variability of catalyst particle and charge size,
(15) Rapid approach to steady state.

Cooke's requirements narrow down the possibilities to gradientless reactors, loop or recycle reactors, or CSTRs, in general.

To the preceding requirements Bennett et al. [1] added the need for:

(16) Low empty-to-catalyst-filled volume ratio.

They found this important for transient studies, but it is also important to minimize homogeneous reactions if a possibility for such reactions exists.

Nelles et al. [19] define in a great detail some of the same requirements mentioned before and add a few more practical criteria:

(17) The catalyst volume should be large enough to hold a representative sample.

(18) It should be possible to test small, less than 0.5 mm, catalyst particles at low gas densities. (We may add that it should be possible to test commercial size catalysts too.)

(19) The construction material should be inert.

(20) Opening and closing as well as maintenance and control of the unit should be problem-free.

From all these requirements it is easy to conclude that the ideal reactor that can handle all reactions under all conditions does not exist. For individual reactions, or for a group of similar reactions, not all the requirements are equally important. In such cases it should be possible to select a reactor that has most of the important attributes.

In the discussion of the various reactors that follows, reference will be made to these numbered requirements as to how well the mentioned reactors satisfy them.

B.   SCALE-DOWN TO LABORATORY REACTORS

Almost a half-century ago, Damköler [20] investigated the possibility of a change in scale for chemical reactors on the basis of the theory of similarity. He concluded that, for tubular reactors that have temperature, concentration, and flow gradients, complete and simultaneous similarity of geometric, mechanical, thermal, and chemical properties is possible only if a single, well-defined reaction is occurring. This is valid for empty homogeneous reactors. For a packed bed catalytic reactor the similarity holds only for laminar flows. These cases are the least interesting industrial applications.

In the example shown in Table I, it is illustrated that, if the geometric similarity of the reactor body is given up, then not only similarity but complete identity can be achieved for the catalyst and its surroundings. In the following example is the hydrogenation of crotonaldehyde to butanol.

$$CH_3CH=CHCHO + 2H_2 \rightarrow CH_3CH_2CH_2CH_2OH$$

which for the conditions selected here is practically irreversible, and for simplification of the example will be considered as such.

The first column in Table I, represents a commercial reactor, and the conditions listed can be considered given. A laboratory reactor has to be devised based on these conditions. The second column represents a single tube identical to one of the 4000 tubes of the commercial reactor. This tube, if the heat transfer side is properly designed, can behave identically to an average tube in the large unit. But this is more of a pilot plant reactor than a laboratory unit because of its size and feed requirements. It gives only the overall performance results unless sample taps are built at equal distances and a thermowell with multiple sensors is built in. In this case it rapidly loses its close identity with the average tube and becomes complicated.

The third column represents an internal recycle reactor where, using a 1 million scale-down factor, all the important flow, thermal, and chemical criteria identities can be maintained for the immediate surroundings of the catalyst. From the operating conditions given at the bottom of the table, holding the rpm constant and changing the fresh feed rate, various conversions can be achieved without changing the mass velocity.

The alternate possibility of building a laboratory tubular reactor that is shorter and smaller in diameter is also permissible, but only for slow and slightly exothermic reactions where smaller catalyst particles also can be used. This would not give a scaleable result for the crotonaldehyde example at the high reaction and heat release rates where pore diffusion limitation can also be expected.

In the petrochemical industry close to 80% of reactions are oxidations and hydrogenations, and consequently they are very exothermic. In addition economy requires fast and selective reactions. Fortunately these can be

TABLE I

Scale-down of Hydrogenation Unit for Crotonaldehyde

	1:1	1:4000	1:1 × 10⁶
Scale-down ratio	$1:1$	$1:4000$	$1:1 \times 10^6$
Feed rate, aldehyde			
lb/yr	200×10^6	50×10^3	200
lb/hr[h]	33.3×10^3	8.33	33.3×10^{-3}
lb mol/hr	462	0.116	0.462×10^{-3}
lb mol, hydrogen[i]	4620	1.16	4.62×10^{-3}
Total feed			
SCFH	1.82×10^6	455	1.82
g mol/hr	2.31×10^6	577	2.31
Tubes			
Number	4000	1	1
i.d. (in.)	1.12	1.12	2.00
Length (in.)	420	420	0.53
Length (ft)	35	35	0.044
Catalyst volume (ft³)	970	0.24	0.97×10^{-3}
Catalyst volume (m³)	27.5	6.9×10^{-3}	27.5×10^{-6}
Pressure drop (atm)	3.8–2.1	3.8–2.1	0.01–0.005

The following conditions will be identical on all three scales at the two different P and T values:

Pressure (psig)	500	300
Temperature		
K	440	490
°C	167	217
°F	333	423
Density, ρ (kg/m³)	8.39	4.63 $\rho_0 = 0.386$
Superficial linear velocity (m/sec)	0.26	0.47
Mass velocity, G (kg/m² sec)	2.18	2.18
Reynolds no., $d_p G/\mu$	1150	932
GHSV (hr⁻¹)	1876	1876

Operating conditions for the recycle reactor:

Flow cross section for 2-in. basket	$A = (2.54)^2(3.14) = 20.27 \text{ cm}^2 = 2.03 \times 10^{-3} \text{ m}^2$
Recycle flow	$F_0 = GA/\rho_0 = 0.0115 \text{ m}^3/\text{sec} = 1457 \text{ SCFH}$
Recycle ratio	$n = 1457/1.83 = 800$
Rpm required	650 at 500 psig and 1150 at 300 psig

[a] Thermowell connection.
[b] Inlet connection.
[c] Outlet connection.
[d] Catalyst basket.
[e] Draft tube.
[f] Impeller.
[g] Magnedrive assembly.
[h] Based on 7500 hr/yr.
[i] 10:1 hydrogen/aldehyde ratio.

studied nowadays in gradientless reactors. The slightly exothermic reactions and many endothermic processes of the petroleum industry still can use various tubular reactors, as will be shown later.

IV. Fluid Bed Reactors

A. FINE SOLID CATALYST IN A SUSPENDED PHASE

Standard agitated autoclaves are the most used reactors for catalysts slurried in reacting liquids. Frequently a gaseous reactant is added as H_2 in hydrogenations in organic chemistry, or the liquid is only a solvent and the reactants are added entirely in gaseous form, as in polymerizations [21]. These tests are made in a semibatch (or semicontinuous) mode, in the sense that product is not removed, whereas the reactant, or one of the reactants, is added continuously.

The potential advantage of running these reactions in CSTRs is recognized widely, but the task of removing the product without removing the fine catalyst slurry is always a discouraging problem. The other possibility of feeding and removing fine solids, especially from a high-pressure reactor, just complicates the problem even more. In large-scale production plants and even in larger pilot plants these feeding and discharging problems of solids can be solved at considerable cost, but they are an operational handicap in small units, in the sense that reliable, continuous operation cannot be maintained for a significant length of time.

Batch or semibatch operation also has the disadvantage of having ill-defined initial conditions from the point of view of the evaluation and mathematical interpretation of measurement results. A variation of the reactor developed for batch-type testing of a particulate catalyst in a two-phase system, the Harshaw reactor [22] can be used. In this reactor the catalyst is held in a basket above the liquid phase until the initial temperature and pressure are established, and then the catalyst in the falling basket reactor drops into the agitated liquid phase.

Agitated autoclave reactors operated in the batch or semibatch mode are far removed from ideal reactors. Yet they can be used successfully, especially if the reaction is not very fast or the catalyst is not very active. With very active catalysts mass transfer usually limits the rate, and in such cases the catalyst in the large-scale unit can behave differently than in the laboratory reactor. Both conversion and selectivity can be significantly impaired in large units if equal performance is expected at the same temperature and pressure on a unit volume basis. Testing with various particle sizes and under various agitating conditions can shed some light on the problem. The

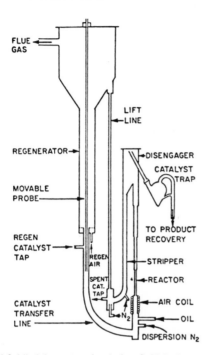

Fig. 1. Atlantic Richfield's laboratory circulating fluid bed reactor. (Courtesy of ARCO Petroleum Co. Harvey, Il.)

scale-up of such results should always take into account the mass transfer differences with increasing sizes.

B. GAS–SOLID FLUID BED REACTORS

These reactors are much needed in the catalytic cracking studies of the petroleum industry. The problems previously mentioned exist here also, perhaps somewhat less severely, and therefore more attempts have been made to overcome them. One of the most successful models is the Arco laboratory version of a fluid catalytic cracker [23]. In this unit some of the problems are alleviated by building the reactor somewhat larger, e.g., for holding 5 lb of catalyst, therefore it is a smaller pilot plant or a so-called bench-scale unit rather than a true laboratory unit. A detailed description and operating characteristics can be found in the literature cited, with comparisons to other standard testing methods. The unit can be purchased under license from Arco and is shown on Fig. 1.

In the Arco unit much lower mass velocities are used than in production reactors, and as such it fails the first requirement. This may not be detrimental, since both reactor and regenerator operate adiabatically and the relative velocity between catalyst particles and gas cannot be very different in laboratory and production size, hence heat and mass transfer between catalyst and gas can be the same. The unit reaches a steady state in a surprisingly short time, permitting two test runs to be completed in a single 8-hr shift. Inspite of the success of this reactor it still puts a high demand for control and operational manipulation on the user. Therefore, other simpler methods were sought for standard testing duty, and these were found in microreactors.

V. Tubular Fixed Bed Reactors

A. MICROREACTORS

1. Reactors for Microactivity Test

The previously discussed Arco reactors, useful as it is, is still too large and too demanding to operate routinely. The oil industry searched for and settled on a standard method for testing fluid cracking catalyst by a microactivity test which has the ASTM designation D3907-80. Details can be found in the corresponding ASTM booklet [24].

The ASTM testing procedure is: "3.1. A sample of cracking catalyst in a fixed bed reactor is contacted with gas oil, an ASTM standard feed. Cracked liquid products are analyzed for unconverted material. Conversion is the difference between weight of feed and unconverted product. 3.2. The standardized conversion is obtained from the measured conversion and the correlation between ASTM reference catalysts and their measured conversions."

As can be seen in the microactivity test a catalyst is tested in a fixed bed reactor in the laboratory to predict its performance in a commercial fluid bed reactor. This can be done only because enormous practical experience exists that has accumulated throughout several decades in several hundreds of reactors both in production and in laboratories. The standard states, "4.1. The microactivity test provides data to assess the *relative* performance of fluid cracking catalysts." Then the standard emphasizes that all operations such as catalyst pretreatment and analytical techniques have to be done as standardized to obtain meaningful relative results.

The microactivity test excells in requirement 4; i.e., it uses small quantities of catalyst, only 4 g, and a feed of 1.33 g in 75 sec, so it is very fast. The test fails almost all the other requirements, and its empirical usefulness is strictly limited to one well-known technology, for an endothermic reaction and one very limited type of catalyst.

2. Small Tubular Reactors for Exploratory Studies

The reactor built for this purpose is usually a $\frac{1}{4}$ in. (6 mm) tube that holds a few cubic centimeters of catalyst. The tube is much longer than needed for the catalyst volume to provide a surface for preheating and to minimize temperature losses at the discharge end. The tube can be bent into a U shape and immersed in a fluidized sand bath, or it can be straight and placed inside a tubular furnace in a temperature-equalizing bronze block. Thermocouples are usually inserted at the ends of the catalyst bed internally, since not much space is available. Outside wall temperature and bath temperature measurements are more commonly used, with the assumption that internal temperatures cannot be very different. This may be a good approximation for slow and not too hot reactions but certainly fails for very hot and fast processes. Good heat transfer on the outside is essential but not sufficient because at low flow rates the inside film coefficient of the reacting stream limits the heat transfer. This is the case not only between the fluid and inner tube wall but also between catalyst particles and the streaming fluid. This is why these reactors frequently exhibit ignition – extinction phenomena and nonreproducibility of results. Laboratory research workers untrained in the field of thermal stability of reactors usually observe that the rate is not a continuous function of the temperature, as the Arrhenius relationship predicts, but that a definite minimum temperature is required to start the reaction [8, 9].

3. Pulse Reactors

Small tubular reactors can be made even smaller and connected directly to the feed line of gas chromatographs and used as pulse reactors. Their usefulness is greatly reduced by the transient operation, because the catalyst is never in steady state with regard to the flowing fluids components in an adsorption – desorption relationship. Results from pulse tests are very questionable, and the interpretation of these results is difficult. Their significance is limited to screening out completely inert or inactive catalysts from active ones, but no quantitative judgment of activity or selectivity can be rendered. Some semiquantitative studies can be made in pulse reactors by repeated pulses for the measurement of adsorbed species and adsorbing poisons.

The pulse reactor, because of its size and operation, is so far removed from the industrial catalytic reactor operations that for practical purposes it is best to avoid its use. It fails so many of the stated criteria that is not worth elaborating on. Choudhary and Doraiswamy [25] review pulse reactors from the academic point of view.

B.   BENCH-SCALE TUBULAR REACTORS

1.   Tube-in-Furnace Reactors

These were the main tools of catalytic studies in the 1950s. Hougen [26] describes several successful models. Most of these reactors have a 25-mm (1-in.) reactor tube which holds 50–100 cm³ catalyst. The same elaborate care is used in adding bronze block liners to smooth out temperature unevenness and in using also multizone furnaces for the same reason. Because of their larger size as compared to the previously discussed micro-reactors, the bronze liner and multizone heating do not help as much as for smaller microreactors. Mass velocities are still much smaller than in production reactors, and Reynolds numbers based on particle diameter are frequently much less than 100 and therefore the flow is in the laminar range. Consequently flow is not similar to that in commercial reactors, and heat and mass transfer are much poorer. These reactors were, and unfortunately still are, used in a few laboratories for process studies on heterogeneous catalysis and frequently with the disastrous results warned against by Carberry [27].

The longer versions holding more (up to 500 cm³) catalyst, especially in long and smaller-diameter tubes, have limited usefulness in the study of high-temperature endothermic reactions such as butane dehydrogenation. For exothermic reactions very small-diameter tubes with a single string of catalyst, sometimes even diluted with dummy carriers between every catalyst particle, make this reactor useful but not very advantageous. All the mentioned precautions make the operating mode of these reactors very different from that of large-scale reactors. In these reactors both the first and second criteria are violated, as well as many others. Therefore, they are obsolete and can be replaced by better reactors.

2.   Thermosiphon Reactors

Figure 2 shows a tubular reactor that has a thermosiphon temperature control system. The reaction is conducted in the vertical stainless steel tube which can have various diameters, $\frac{1}{2}$ in. being the preferred size. If used for fixed bed catalytic studies, it can be charged with a single string of catalytic particles just a bit smaller than the tube, e.g., $\frac{3}{8}$-in. particles in a $\frac{1}{2}$-in. tube. With a smaller catalyst, a tube with an inside diameter of up to three to four particle diameters can be used. With such catalyst charges and a reasonably high Reynolds number—above 500, based on particle diameter—this reactor approximates fairly well the performance of an ideal isothermal plug flow reactor for all but the most exothermic reactions.

The temperature inside the tube is difficult to measure, and with a single

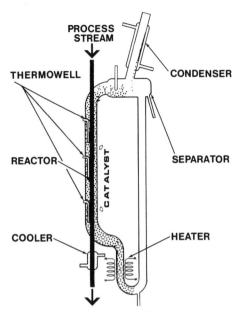

Fig. 2. Thermosiphon jacketed reactor.

string of catalyst one has to be satisfied with measuring it at the end of the bed. This can be accomplished by using a thermocouple inserted from the bottom. This thermocouple can also serve as the catalyst retainer, or bed support. The temperature in the jacket can be kept constant by controlling the inert gas pressure at the top of the reflux cooler. The downcomer line is filled with liquid, and the reactor jacket has vapor bubbles in the liquid and serves as a riser. The coolant is partially evaporated in the boiler; more is evaporated by absorbing the exothermic reaction heat in the jacket. The vapor–liquid mixture is lifted by density difference, thus creating circulation. Since the jacket side heat transfer is in the boiling mode, it is very high and the tube (reaction) side film coefficient is controlling. The thermosiphon circulation rate can be as high as 10 to 15 times the coolant evaporation rate. This, in turn, eliminates any significant temperature difference, and the jacket is maintained under isothermal conditions. In this case the constant wall temperature assumption is satisfied. During heating up of the thermosiphon, the bottom can be 20–30°C hotter, and the start of the circulation can be established from the observation that the temperature difference between the top and bottom jacket temperature is diminishing.

    The ½-in. i.d. reactor tube has the advantage of a relatively small wall thickness, even at higher pressures, hence permits better heat transfer. The

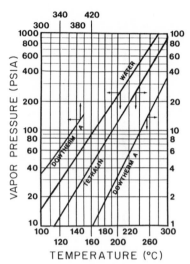

Fig. 3. Vapor pressure–temperature relationship for coolants.

heating or cooling jacket can also be designed for high pressure. The 400-psig rating satisfies most needs, although higher jacket pressures can be accommodated too.

Figure 3 represents the vapor–pressure–temperature relationship for three coolants: water, tetralin, and Dowtherm A. It is interesting to note that, at high jacket pressures, a single Celsius temperature change can cause more than 1 psi pressure difference. Since pressure can be controlled much better than this difference, excellent temperature control can be achieved in the jacket by maintaining constant pressure over the boiling liquid.

3. Liquid-Cooled Reactors

As the name implies, these reactors are mostly used for the study of exothermic reactions, although they can be applied for endothermic reactions also.

Figure 4 shows a liquid-jacketed tubular reactor. In this arrangement, in contrast to the previous approach, the coolant is kept from evaporating by maintaining it under an inert gas pressure higher than its vapor pressure. A centrifugal pump is used to achieve high circulation rates. Besides the previously mentioned three coolants (water, tetralin, and Dowtherm A), other nonvolatile heat transfer oils as well as molten salts or molten metals can be used. These coolants are advantageous primarily at higher temperatures where, even if exothermic reactions are conducted in laboratory reactors, the heat losses are usually more than the reaction heat generated. A

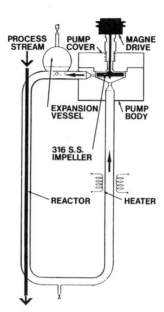

Fig. 4. Liquid-cooled reactor.

simple temperature controller, therefore, can be used to keep the electric heater adjusted by sensing with a thermocouple immersed in the return line.

At a high recirculation rate of the liquid coolant, the constant wall temperature can be approximated again, but not as well as with boiling-type cooling. On the other hand, this type of cooling permits the study of increasing or decreasing temperature profiles in the jacket and the influence of these on the inner temperature profile, reactor performance, and stability. For this type of study a reactor tube is needed that is large enough to accommodate an inner thermowell holding a multiple thermocouple assembly.

Recirculation of the nonboiling liquids can be achieved by bubbling inert gas through the liquid in the reactor jacket. This is less practical with fluids that have significant vapor pressure, because the jacket still has to be under pressure and a large condenser has to be installed to condense out the liquid from the gas saturated with vapors at the jacket temperature. It is more useful with molten metals and salts.

For the design details of the inside of the reactor tube, the same considerations apply as for a thermosiphon-controlled reactor.

Although fluidized sand or alumina can also be used for these somewhat larger reactors, the size makes the jacket design a problem in itself, hence these reactors are seldom used. An advantage of the jacketed reactor is that several, usually four, parallel reactor tubes can be placed in the same jacket and operated at one common jacket temperature, but otherwise under different conditions if needed. This type of arrangement saves times and space in longer-lasting catalyst life studies.

Jacketed tubular reactors come close but still cannot reproduce industrial conditions as required by the first and second conditions. Liquid-cooled reactors like thermosiphon reactors can be used for all but the most exothermic and fast oxidation reactions. These reactors are commercially available from Design Technology, Inc., Pittsburgh, Pennsylvania.

VI. Gradientless Reactors

In gradientless reactors we try to carry out the catalytic reaction under highly, even if not completely, uniform conditions of temperature and concentrations. The reason for this is that, if achieved, the subsequent mathematical analysis and kinetic interpretation will be simpler to perform and the results can be used more reliably. The many ways of approximating gradientless operating conditions in laboratory reactors will be discussed next.

A. THE DIFFERENTIAL REACTOR

In a differential reactor the concentration change, i.e., the conversions increase, is kept so low that the effect of the concentration and temperature changes can be neglected. On the other hand the concentration change has to be quantitatively known because this, multipled by the flow rate and divided by the catalyst quantity, measures the reaction rate as

$$dC/dt \simeq \Delta C/\Delta t = r = \frac{C_{in} - C_{out}}{W/F} \qquad \frac{mol/m^3}{(kg)(m^3/sec)}.$$

This rate, measured the previous way, has to be correlated with the temperature and concentration as in the following simple power law rate expression:

$$r = A \exp(-E/RT)[(C_{in} + C_{out})/2]^n.$$

There are two contradictory requirements here. The first is to keep the difference between C_{in} and C_{out} as small as possible so that it can be neglected. The second is to analyze these two only very slightly different concentrations with such precision that the difference will be significantly greater than the measurement error. The second need is for calculation of the rate of reaction, as shown in the first equation of this section.

This is an obviously difficult task, and it is rarely possible to satisfy both requirements reasonably and simultaneously. This difficulty is compounded by the need to use a preconverter to achieve the various conversion levels where the additional incremental increase in conversion can be measured. The alternate way is to feed the reactor various amounts of products in addition to the starting material.

B.   THE LOOP OR RECYCLE REACTOR

In a differential reactor the product stream differs from the feed only very slightly, therefore the addition of products to the feed stream can be avoided if one recycles most of the product stream. The feed can be made up mostly from the recycle stream with just enough starting materials added to replace that which was converted in the reaction and blown off in the discharge stream. This is the basis of loop or recycle reactors.

In Fig. 5, the material balance of a recycle reactor is illustrated by a Sankey diagram where the widths of bands are proportional to the flowing masses. If all the chemical change is effected by W kilograms of catalyst, an outer and an inner material balance can be written. These two balances have to be equal to each other and account for the change. From this balance it follows that the difficulties associated with errors in chemical analysis of the small concentration difference $C_1 - C$ are eliminated by measuring $C_0 - C$, which is $(C_0 - C)/(C_1 - C) = n + 1$ times larger than $C_1 - C$.

The internal balance in Fig. 5 represents the material balance for a

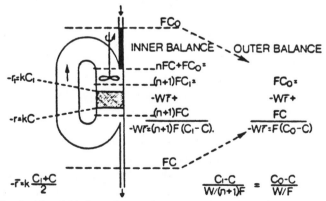

Fig. 5. Material balance for recycle reactors. (Reproduced from Berty [7].)

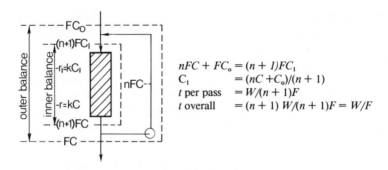

Along with the diagram, the following equations appear:

$$nFC + FC_o = (n + 1)FC_1$$
$$C_1 = (nC + C_o)/(n + 1)$$
$$t \text{ per pass} = W/(n + 1)F$$
$$t \text{ overall} = (n + 1)\,W/(n + 1)F = W/F$$

Integrating	Averaging
$-(n + 1)F\,dC/dW = kC,$	$-r = (n + 1)F(C_1 - C)/W$
$-\int_{C_1}^{C} dC/C = k/(n + 1)F \int_0^{W} dW$	from the identity of inner and outer balances:
$-\ln(C/C_1) = kW/(n + 1)F$	$-r = F(C_o - C)/W$
	$-r = k(C_1 + C)/2$
$C = C_1 e^{-kW/(n + 1)F}$	$F(C_o - C)/W = k(C_1 + C)/2$

Fig. 6. Recycle reactor—integration and averaging between inner limits. (Reproduced from Berty [7].)

differential reactor where the preconverted feed is made from the recycle and a small fresh feed.

In Fig. 6, the performance of a recycle reactor is calculated in two different ways. The integration for the inner limits is valid at any value of n, while the averaging is justified only for reasonably large values of n. In Fig. 7, an ideal plug flow tubular reactor and an ideal CSTR are shown for two extreme cases for the recycle reactor, when n tends to zero or to infinity, respectively.

The conclusion to remember is that, in recycle reactors at high recycle ratios, all the reaction proceeds at, or very closely to discharge conditions like those in an ideal, perfectly stirred CSTR. Therefore, good recycle reactors and good CSTRs can be used for the same purposes equally well, contrary to some misconceptions [4]. The only difference is that, with the well-defined flow path of recycle reactors, the "goodness" can be ascertained, whereas with CSTRs, like rotating basket reactors, the mixing can be perfect yet the contact between the catalyst and gas still poor or uncertain.

The limit condition $n \rightarrow \infty$ is approximated experimentally when the concentration difference before and after the catalyst drops to the error limit of either the chemical analysis or the observable effect of concentration difference on the rate of reaction. This is usually achieved at about 20–25 recycle ratios, although for a higher precision measurement with concentrated systems without a diluent higher values are needed.

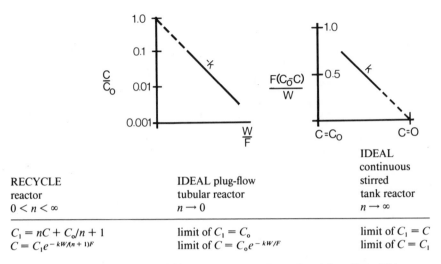

Fig. 7. Extreme conditions for recycle reactors. (Reproduced from Berty [7].)

The other limit condition $n \rightarrow 0$ leads to the tubular reactor, which was discussed previously. The various practical construction forms of gradientless reactors will now be discussed.

C. CONSTRUCTION CONCEPTS FOR GRADIENTLESS REACTORS

Jankowski *et al.* [5] discuss in detail the great variety of gradientless reactors proposed by several authors. All of these reactors can be placed in a few general categories: (1) moving catalyst basket reactors, (2) external recycle reactors, and (3) internal recycle reactors.

1. Moving Catalyst Basket Reactors

Moving catalyst basket reactors all fail the first requirement because the flow regime is ill-defined and the contact between catalyst and gas can be poor even if the tenth requirement, namely, well-mixed conditions, is satisfied. Perhaps the most successful representative of this category is the Carberry reactor [27, 28]. Even in this model only a single layer of catalyst can be charged in the cruciform catalyst basket because the fluid flows in a radial direction outward and does not penetrate much of the catalyst basket. This reactor is shown on the right in Fig. 8.

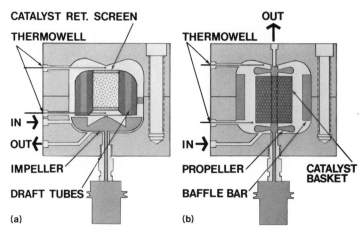

Fig. 8. Berty (a) and Carberry (b) reactors. (Courtesy of Autoclave Engineers, Inc. Erie, PA.)

2. External Recycle Reactors

The major difficulty with these reactors is in the outside recycle pump, especially at high temperatures. Reciprocating pumps require seal rings, and these cannot take the high temperature needed for most reactions. If the recycle gas is cooled down before entering the compressor, it has to be reheated before it enters the reactor again. This makes them fail the ninth test involving construction difficulty and costs.

Single-stage blowers or tubines usually cannot generate enough of a pressure difference to overcome the added resistance of the recycle pipes. Multistage blowers are very expensive and add even more to the gas-filled space (item 16), and this in turn makes them approach a steady state slowly (item 15). In addition some components may condense out in the cooler, especially with high-boiling materials and/or high pressures.

In summary external recycle reactors are expensive and/or their usefulness is limited. For simple chemical systems where no condensation can occur and neither high pressure nor high temperature is needed they can be practical. For example Carberry [29] prefers an external recycle reactor for the study of CO oxidation over a spinning basket reactor.

3. Internal Recycle Reactors

These are the most successful types of reactors presently available. The internal reciprocating plunger types, for example, that of Nelles et al. [19], do not provide a steady uniform flow. Of those operating with rotating blowers or turbines the best known are those of Garanin et al. [30], Brown and Bennett [31], Livbjerg and Villadsen [32], Boag et al. [33], Berty et al.

Fig. 9. Berty reactor for exploratory studies. (Courtesy of Design Technology, Inc. Pittsburgh, PA.)

[18], and Berty [17]. These reactors all work on very similar basic principles and will be discussed based on the example of the Berty reactor, which is manufactured commercially and more than 250 are in operation around the world.

The Berty reactor is shown on the left in Fig. 8. This reactor satisfies most of the requirements, but even it fails the ninth (cost), sixteenth, (empty-to-catalyst-filled volume ratio), and twentieth (quick opening and maintenance) tests. Carberry and Berty reactors are commercially available from Autoclave Engineers, Inc., Erie, Pennsylvania.

The latest version of the Berty reactor is shown in Fig. 9. This 2-in. model was developed for quick exploratory studies on small samples of catalysts. The maximum catalyst sample volume is 15 cm³, and the reactor has a single-bolt-operated, quick opening closure. Besides that smaller quantities of catalyst are easier to make in a research laboratory, the smaller size also makes the use of expensive starting materials or labeled intermediates more efficient.

Another 4-in. model is under development for use in low-pressure and/or high-temperature studies. In such cases very high shaft speeds are needed to generate a great enough pressure difference for reasonably high mass velocities.

Both new models were developed at Design Technology, Inc., from which they are available commercially. The operational characteristics of all Berty reactors are described in Berty [17], and their use in catalyst testing in Berty [7]. Typical uses for ethylene oxide catalyst testing are described in Bhasin [34].

Internal recycle reactors are easy to run with minimum control or automation. Complete automation with computer control and on-line data evaluation and reduction is also possible [35–37].

VII. Conclusion

Although there are many catalytic reactors for various uses, the role of gradientless reactors is expanding rapidly. Among these, internal recycle reactors are the most popular. These reactors can solve many, but not all, problems. Tubular reactors will retain significance in testing the lifetime of slowly deactivating catalysts. Larger tubular reactors will be used in laboratory and in pilot plant sizes to test predictions based on gradientless reactors.

References

1. C. O. Bennett, M. B. Cutlip, and C. C. Yang, *Chem. Eng. Sci.* **27**, 2255–2264 (1972).
2. V. W. Weekman, *AIChE J.* **20**, 5, 833–840 (1974).
3. L. K. Doraiswamy, and D. G. Tajbl, *Cat. Rev. Sci. Eng.* **10**(2), 177–219 (1974).

4. A. M. R. Difford, and M. S. Spencer, *Chem. Eng. Prog.* **71**(1), 31–33 (1975).
5. H. Jankowski, J. Nelles, R. Adler, B. Kubias, and C. Salzer, *Chem. Tech.* (*Berlin*) **30**(9), 441–446 (1978).
6. C. G. Cooke, *CHEMSA,* **5**(11), 175–176 (1979).
7. J. M. Berty, *Catal. Rev. Sci. Eng.* **20**, 1, 75–96 (1979).
8. D. D. Perlmutter, "Stability of Chemical Reactors." Prentice-Hall, Englewood Cliffs, New Jersey (1972).
9. J. M. Berty, J. P. Lenczyk, and S. M. Shah, *AIChE J.* **28**, 6, 914–922 (1983).
10. J. J. Carberry, and A. A. Kulkarni, *J. Catal.* **3**, 141 (1973).
11. D. E. Mears, *Ind. Eng. Chem. Proc. Des. Dev.* **10**(4), 541–547 (1971).
12. J. J. Carberry, *AIChE J.* **6**, 460 (1960).
13. D. F. Boucher, and G. E. Alvez, *Chem. Eng. Prog.* **55**(9), 55–64 (1959).
14. D. F. Boucher, and G. E. Alvez, *Chem. Eng. Prog.* **59**(8) 75–83 (1963).
15. E. W. Thiele, *Ind. Eng. Chem.* **31**, 916 (1939).
16. P. B. Weisz, and C. D. Prater, *Adv. Catal.* **6**, 143 (1954).
17. J. M. Berty, *Chem. Eng. Prog.* **70**, 5, 78–83 (1974).
18. J. M. Berty, J. O. Hambrick, T. R. Malone, and D. S. Ullock, *Preprint 42E of the 64th Nat. Meeting of AIChE,* New Orleans, Louisiana, March 16–20 (1969).
19. J. Nelles, H. Jankowski, R. Adler, B. Kubias, and C. Salzer, *Chem. Tech.* (*Berlin*) **30**(1), 555–559 (1978).
20. G. Damköhler, *Z. Elektrochem.* **42**, 12, 846–862 (1936).
21. R. F. Gold, S. F. Gelman, J. R. Doonan, and G. G. Arzoumanidis, *Chem. Eng.* Jan. 31, 119–121 (1977).
22. W. R. Alcorn, G. E. Elliott, and L. A. Cullo, *Preprint ACS Fuel Chem. Div. ACS Nat. Meeting,* March (1978).
23. S. J. Wachtel, L. A. Baillie, R. L. Foster, and H. E. Jacobs, *Oil Gas J.* April 10, 104–107 (1972).
24. ASTM Standards on Catalyst, PCN 06–432080–12 American Society for Testing Materials, Philadelphia, Pennsylvania (1980).
25. V. R. Choudhary, and L. K. Doraiswamy, *Ind. Eng. Chem. Prod. Res. Dev.* **10**, 219 (1971).
26. O. A. Hougen, *AIChE Monogr. Ser.,* **47**, 50–53 (1951).
27. J. J. Carberry, *Ind. Eng. Chem.* **56**(11), 39 (1964).
28. D. G. Tajbl, J. B. Simons, and J. J. Carberry, *Ind. Eng. Chem. Fundam.* **5**, 171 (1966).
29. J. J. Carberry, S. C. Paspek, and A. Varma, *Chem. Eng. Educ.* **14**(2), 78 (1980).
30. V. I. Garanin, U. M. Kurkchi, and Kh. M. Minachev, *Kinet. Katal.* **8**(3), 701–703 (1967).
31. C. E. Brown, and C. O. Bennett, *AIChE J.* **16**, 817 (1970).
32. H. Libjerg, and J. Villadsen, *Chem. Eng. Sci.* **26**, 1495–1503 (1971).
33. I. F. Boag, D. W. Bacon, and J. Downic, *J. Catal.* **38**, 375 (1975).
34. M. M. Bhasin, P. C. Ellgen, and C. D. Hendrix, UK Patent Application GB2043481 A (1980).
35. R. D. Dean, and N. B. Angelo, *Chem. Eng. Prog.* **67**(10), 59–68 (1971).
36. C. D. Gregory, and F. G. Young, *Chem. Eng. Prog.* **74**(5), 44–48 (1979).
37. F. P. Larmon, M. M. Gilbert, and R. R. Dean, *2nd World Congr. Chem. Eng.* Paper No. 10.03.02 (1981).

CHAPTER 4

Catalytic Hydrotreating in Petroleum Refining

DONALD C. McCULLOCH

Worldwide Catalyst Department
American Cyanamid Company
Wayne, New Jersey

I. Introduction

A. DEFINITION OF HYDROTREATING

In the petroleum refining industry, catalytic hydrogen treating (or hydrogen processing) can be defined as the contacting of petroleum feedstocks with hydrogen, in the presence of a suitable catalyst and under suitable operating conditions, to effect conversion to lower molecular weight hydrocarbons, to prepare the feedstocks for further conversion downstream, and/or to improve the quality of finished products.

Hydrogen treating can be separated into several categories. Aalund and Cantrell [1] define hydrocracking as processes where 50% or more of the feed is reduced in molecular size, hydrorefining as processes where a small amount (up to 10%) of the feed is reduced in molecular size, and hydrotreating as processes where essentially no molecular size reduction occurs. Other process descriptions frequently used are hydrodesulfurization (HDS) and hydrodenitrogenation (HDN).

For the remainder of this chapter, "hydrotreating" will be used as a general term (and the one most commonly used today) to include all these processes except hydrocracking, which will be covered in a subsequent volume.

B. COMMON OBJECTIVES AND APPLICATIONS OF HYDROTREATING

(1) Naphtha (catalytic reformer feed pretreatment)—to remove sulfur, nitrogen, and metals that otherwise would poison downstream noble metal reforming catalysts,

(2)   Kerosene, jet fuel, diesel oil, heating oil—to remove sulfur and to saturate olefinic and aromatic molecules, resulting in improved smoke point, cetane number, diesel index, and storage stability,

(3)   Lube oil—to improve the viscosity index, color and color stability, and storage stability to reduce gum formation, and to decrease the neutralization number,

(4)   Gas oil [fluid catalytic cracker (FCC) feed pretreatment]—to improve FCC yields, reduce FCC catalyst usage and stack emissions, and minimize corrosion by the reduction of feed sulfur, nitrogen, metals and polynuclear aromatics (PNA), and

(5)   Resids (atmospheric and vacuum-reduced crudes)—originally only to provide low-sulfur fuel oils, but more recently also to effect conversion and/or pretreatment for further conversion downstream.

These applications, and others, will be discussed in more detail in Section IV.

C.   EXTENT OF COMMERCIAL USE

Prior to the mid-1950s, the development of hydrotreating processes on a large commercial scale was limited by both the high cost of hydrogen and the lack of knowledge about high-pressure processing. The former problem was greatly alleviated by the rapid growth of catalytic reforming (which generates large amounts of net hydrogen) in the 1950s and 1960s. The latter problem has been reduced significantly as a result of technological advances and, perhaps more importantly, the confidence established over many years of successful commercial experience.

Good data on the extent of the commercial use of various catalytic processes can be found in the *Oil and Gas Journal* refining surveys for the United States and Canada (Aalund and Cantrell [1], plus earlier annual issues) and for the rest of the free world (Auldridge and Cantrell [2] plus earlier annual issues).

Figure 1 plots the recent growth of hydrotreating capacity in the United States from 1976 to 1981, which calculates to an annual average increase of between 6 and 7%. During this period, U.S. hydrotreating capacity as a percentage of crude refining capacity also increased, from 41.4% in January 1976 to 45.4% in January 1981.

Table I lists comparable information on hydrotreating capacities worldwide as of January 1981. Hydrotreating as a percentage of crude capacity remains considerably higher in the United States and Canada than in the rest of the free world, primarily because of more upgrading of feedstocks to improve downstream conversion processing (catalytic reforming, FCC, etc.)

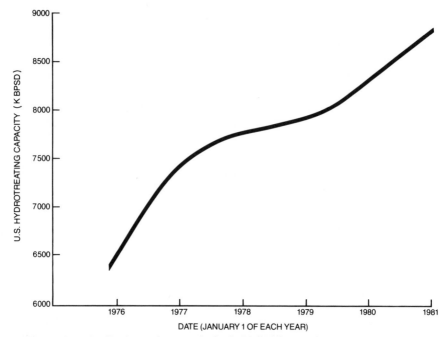

Fig. 1. Growth of hydrotreating capacity in the United States, 1976–1981. Source: *Oil and Gas Journal.*

II. Chemical Reactions

A. DESULFURIZATION

The chemical reactions most frequently of interest in petroleum hydrotreating are those involving the removal of sulfur from hydrocarbon molecules. Figure 2 shows three typical desulfurization reactions and the amount of hydrogen consumed.

Note that H_2 is required not only to react sulfur to hydrogen sulfide but also to saturate (or partially saturate) the desulfurized hydrocarbon. In fact, when cyclic molecules are involved, ring saturation normally takes place before the sulfur (or nitrogen or oxygen) atom is broken off.

Reaction kinetics and mechanisms will not be discussed in much detail in this chapter, since the intended focus is on commercial applications technology. A good reference covering the catalytic chemistry of HDS is Gates *et al.* [3].

TABLE I

Hydrotreating Capacities Worldwide: January 1981[a]

Country	Feed capacity (kBPSD)[b]		Hydrotreating as percentage of crude
	Hydrotreating	Crude	
United States	8,784	19,371	45.4
Canada	981	2,302	42.6
All other free world countries	13,939	50,066[c]	27.8
	23,704	72,739	33.0

[a] Source: *Oil and Gas Journal.*
[b] kBPSD, Barrels per stream day $\times 10^3$.
[c] Calculated as barrels per calendar day (BPCD) $\div$ 0.90, assuming a 90% on-stream factor for crude.

B. DENITROGENATION

Removal of nitrogen is important for many petroleum streams, such as those being pretreated for feed to catalytic reformers, fluid catalytic crackers, and hydrocrackers, especially when feed is derived from high-nitrogen crude oils such as those typically found in California. Nitrogen removal will become even more critical in the future when significant amounts of high-nitrogen synthetic crudes from coal and oil shale must be upgraded by

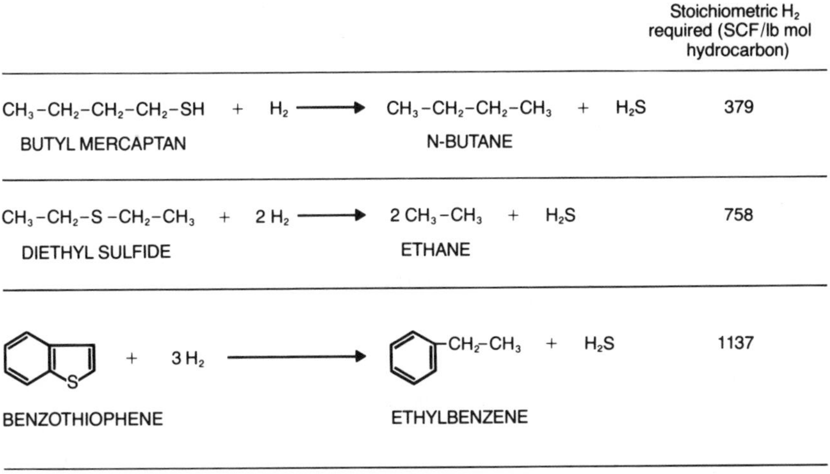

Fig. 2.   Typical desulfurization reactions. SCF, Standard cubic feet at 1 atm and 60°F.

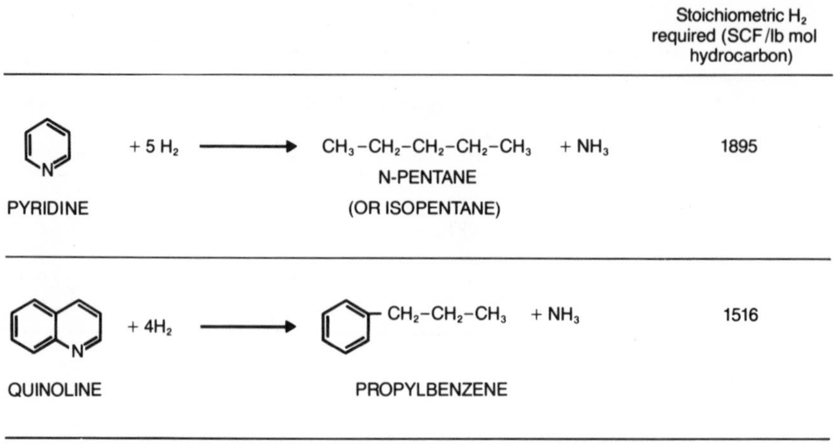

Fig. 3. Typical denitrogenation reactions.

hydrotreating. Two examples of typical denitrogenation reactions are shown in Fig. 3.

C. DEOXYGENATION

Most petroleum crudes do not contain large amounts of oxygen. The pretreater required to remove sulfur and nitrogen generally also removes oxygen adequately under the same operating conditions. High-oxygen synthetic crudes from coal, oil shale, and tar sands will make oxygen removal

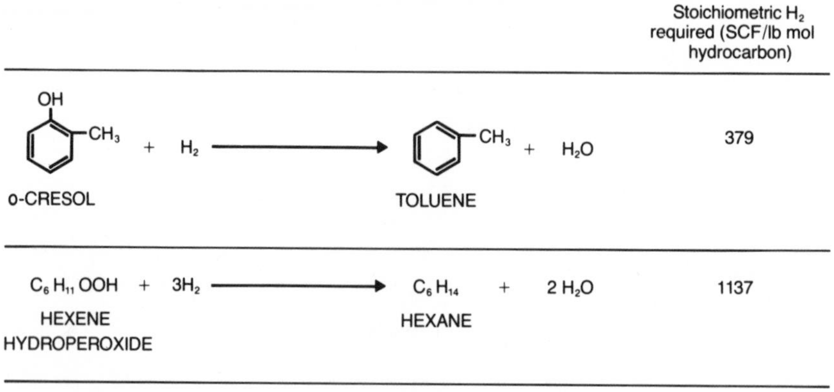

Fig. 4. Typical deoxygenation reactions.

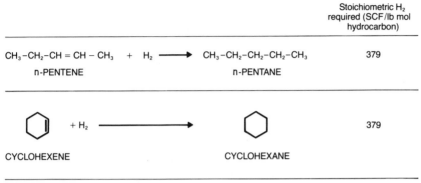

Fig. 5. Typical olefin saturation reactions.

more critical in the future. Examples of deoxygenation reactions are shown in Fig. 4.

D.   OLEFIN SATURATION

Saturation of olefins (and diolefins) is especially important for feedstocks derived from thermal cracking operations such as coking and ethylene manufacture. Such feedstocks are unstable and must be carefully protected from contact with oxygen, which could result in the formation of polymer gums before hydrotreating.

Examples of olefin saturation reactions are shown in Fig. 5. These reactions are highly exothermic, so large amounts of olefins in the feed require special attention to hydrotreater temperature control, frequently including the recycle of hydrotreated product to serve as a heat sink.

E.   AROMATIC SATURATION

Saturation of monoaromatic (benzene) rings sometimes is desirable for improvement of the smoke point, diesel index, etc., for certain middle-distillate products. A significant reduction (say 25% or more) in monoaromatic rings requires fairly severe hydrotreating conditions because the single ring is quite stable. It should also be noted that monoaromatic saturation of any petroleum stream that eventually yields a gasoline fraction is generally considered undesirable because of the resultant decrease in octane.

Figure 6 shows an example of a PNA chain reaction, which typically yields all the products in this chain, the amounts of each varying according

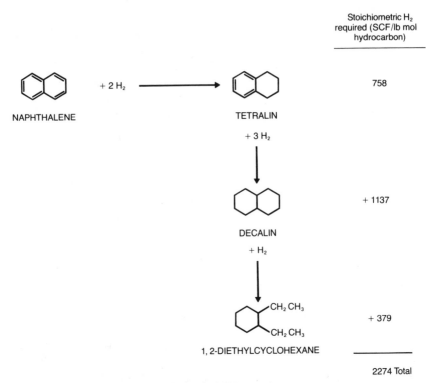

Fig. 6.   Typical PNA reactions.

to the severity of hydrotreating conditions. PNA saturation can consume large amounts of hydrogen.

F.   DEMETALLATION AND ASPHALTENE CONVERSION

Residuum feedstocks, especially those derived from heavy crudes, can contain high levels of asphaltenes and metals. Conversion of these asphaltenes, which are the major contributors to Conradson carbon residue, and removal of these contaminant metals, especially nickel and vanadium, are becoming increasingly important in order to produce feedstocks suitable for downstream FCC or hydrocracking conversion units.

Asphaltenes are among the highest molecular weight components in crude oil. Although their exact structures are not well defined, Fig. 7 gives a general idea of what they look like. As can be seen, they are highly con-

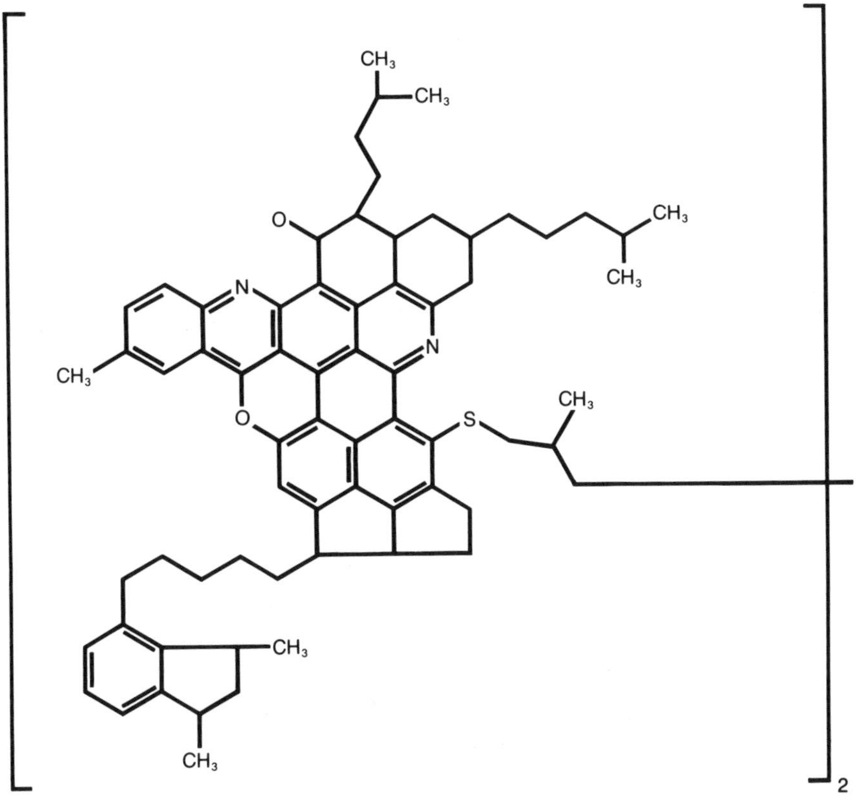

Fig. 7. Hypothetical asphaltene structure.

densed (low hydrogen/carbon ratio) polycyclic compounds generally containing heteroatoms of sulfur, nitrogen, and/or oxygen.

Frequently associated with asphaltenes are complex organic structures called porphyrins. Figure 8 shows two examples of porphyrin-type structures. In each case, the nucleus is a flat ring consisting of four pyrrole nuclei arranged such that nitrogen atoms point inward and form a "cage" in which metal atoms can be chemically trapped [4]. Each nitrogen atom has a free electron pair (shown as two dots) which coordinates with metal atoms to hold them firmly in place. However, over the thousands, even millions, of years that these compounds have spent below the ground, some substitution of different metals has occurred. Because nickel and vanadium porphyrins are especially stable structures, this substitution process could explain why nickel and vanadium are the most common metals found in crude oil.

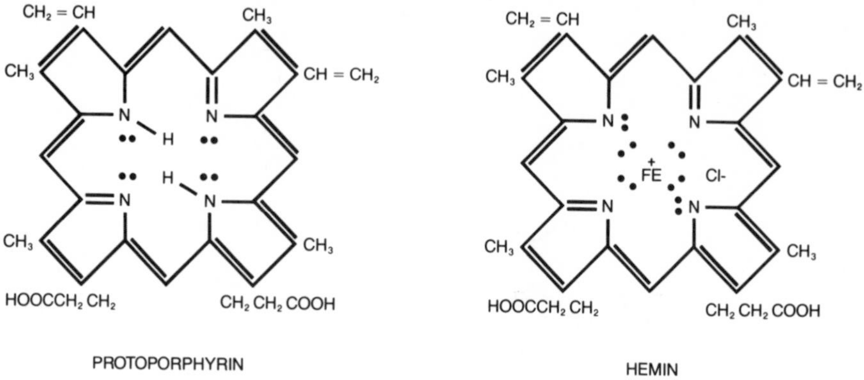

PROTOPORPHYRIN   HEMIN

Fig. 8. Porphyrin-type structures.

III. General Process Description

A. RELATIONSHIP OF HYDROTREATING TO OTHER REFINERY OPERATIONS

Figure 9 depicts the location of four commonly used types of hydrotreating units in relationship to other major refinery process operations. Further comments are as follows:

(1) Naphtha hydrotreater—primary function is to pretreat feed before catalytic reforming or isomerization. Heavy naphtha (180°F+) normally is fed to a catalytic reformer. Light straight-run (SR) naphtha (C_5 to 180°F) commonly is fed to an isomerization unit or blended directly into the gasoline pool.

(2) Distillate hydrotreater—primary function is to reduce sulfur to required specification levels and, in some cases, to saturate aromatics to improve the smoke point, diesel index, etc. Although not shown here, thermally cracked distillate products, such as those from a coker or visbreaker, generally must be hydrotreated also.

(3) Gas oil hydrotreater—primary function is to pretreat feed before fluid catalytic cracking or hydrocracking or to desulfurize gas oil sufficiently to meet heavy fuel oil sulfur specifications after blending with undesulfurized atmospheric or vacuum resid (so-called indirect resid desulfurization).

(4) Resid hydrotreater—primary function always has been to desulfurize sufficiently to meet heavy fuel oil sulfur specifications, but many refiners are now planning to use these units also to effect conversion and/or to

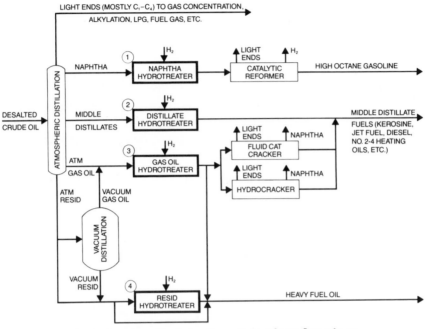

Fig. 9. Location of hydrotreating units in refinery flow scheme.

pretreat resid for further conversion downstream. This kind of higher severity operation must involve demetallation as well as desulfurization.

Three other common hydrotreating applications not specifically depicted in Fig. 9 are upgrading lube oils, specialty solvent hydrogenation, and certain sulfur recovery plant tail gas processes.

B. HYDROGEN SUPPLY REQUIREMENTS

All hydrotreating units, by definition, must consume significant amounts of hydrogen. This quantity varies from only a few SCF/bbl for hydrotreating some low-sulfur SR naphthas to over 1000 SCF/bbl for some high-severity resid units.

Catalytic reforming is the major net producer of hydrogen used by most hydrotreating units today. Hydrogen generated by this process will be quantified in Chapter 5. Large H_2-consuming units, such as hydrocrackers and resid hydrotreaters, frequently require additional H_2 production facilities. This most commonly comes from steam reforming of natural gas, liquid petroleum gas (LPG), or light naphtha. Some refiners are starting to generate H_2 by partial oxidation of residual fuel, which is a less expensive feed.

C. HYDROTREATING UNIT FLOWSHEETS

Figure 10 is a simplified block diagram of a typical commercial hydrotreating unit. All hydrotreating units include a preheat system of some kind, one or more reactors, and a gas–liquid separation system. Several common variations among hydrotreating units are:

(1)   Most units have several feed–effluent heat exchangers, followed by one or two preheat furnaces, depending upon whether the oil feed and H_2 treat gas are preheated together or separately.

(2)   The simplest units, normally for SR low-sulfur naphtha or low-severity distillate hydrotreating, have only one reactor with a single bed of catalyst. The most complex units, such as those for hydrotreating resids, can

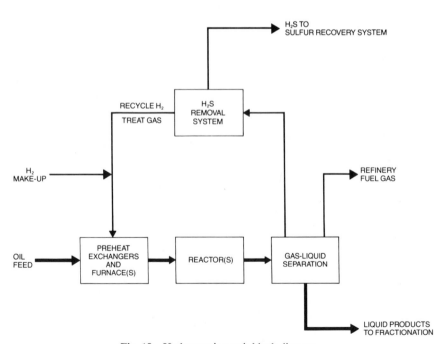

Fig. 10.   Hydrotreating unit block diagram.

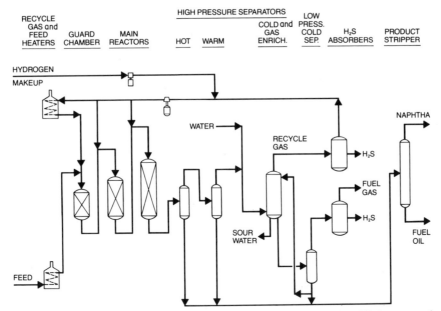

Fig. 11. Unicracking/HDS simplified process flowsheet. Source: Union Oil Company of California.

include several reactors in series and sometimes two complete parallel trains of reactors. When reaction exotherms are high, nonpreheated recycle hydrogen can be introduced separately into the reactors downstream of the first reactor in order to keep temperatures from exceeding the desired maximum levels (called hydrogen quenching).

(3)   Gas–liquid separation can be accomplished simply in a single-stage flash drum, but most units contain two or more separation drums plus some kind of stripping tower. Some units contain as many as four separation stages, two at high pressure and two at low pressure and each with a hot and a cold section. The purpose of these multiple separation steps is to maximize both recycle gas H_2 purity and the removal of light ends (mostly C_1 to C_4) from liquid products.

(4)   Wherever the feed sulfur level is appreciable (say, 0.5 wt% or more), improved catalyst performance (and perhaps unit metallurgy) should justify the installation of a recycle gas scrubber to remove H_2S. The inhibiting effect of H_2S (and NH_3) on hydrotreating reactions will be discussed in Section VI.

As an example of a specific commercial process, Fig. 11 is a simplified flow diagram for Unicracking/HDS, a residual oil hydrotreating process developed and licensed by the Union Oil Company of California. Note that this flowsheet illustrates many of the items mentioned above, such as hydrogen quenching, multistage liquid–gas separation, and H_2S removal. This process also includes a bypassable guard chamber reactor that maximizes the removal of particulate matter and residual salt content from the feed, thus prolonging cycle length by delaying contamination of the main downstream reactors.

D. TYPES OF REACTORS

1. Fixed Bed

The majority of hydrotreating reactors in commercial use today contain fixed beds of catalyst of either downflow or radial flow design. Each of these designs is illustrated in Fig. 12.

DOWNFLOW. As shown, feed (mixed oil and H_2 treat gas) enters at the top of the reactor, flows downward through the catalyst bed, and exits at the bottom. The support balls shown at the top of the bed help to trap any particulate matter entering the reactor, to hold the catalyst bed in place, and to distribute the feed uniformly across the reactor cross-sectional area. The grid and balls at the bottom of the reactor serve to support the catalyst bed. The loading and distribution of catalyst and support balls will be discussed in Section V. When the feed is mixed vapor and liquid, this kind of reactor is commonly called a trickle bed reactor. Downflow reactors can contain multiple beds of catalyst, separated by appropriate grids, support balls, and trays, in order to redistribute flow to achieve better catalyst effectiveness and/or to provide temperature control by H_2 quenching between beds.

RADIAL FLOW. In this design, feed enters at the top of the reactor and then flows more or less horizontally through an annular catalyst bed to a center pipe product collector. Properly sized metal screens surround the center pipe and the outside of the catalyst bed to hold the bed in place. Radial flow reactors can be used only when all feed is vaporized, of course, because otherwise entering gas and liquid would separate quickly by gravity in the annular space both outside and within the catalyst bed. For applications that are 100% vapor phase, radial flow reactors can offer the advantages of a lower pressure drop, because of the large cross-sectional area and short linear flow path through the catalyst bed, and increased resistance to plugging of the catalyst bed by particulate matter carried over in the feed. However, this same short linear flow path can become a big problem if settling of the bed and/or movement of the screens or center pipe results in the opening up of low-resistance flow channels through the catalyst.

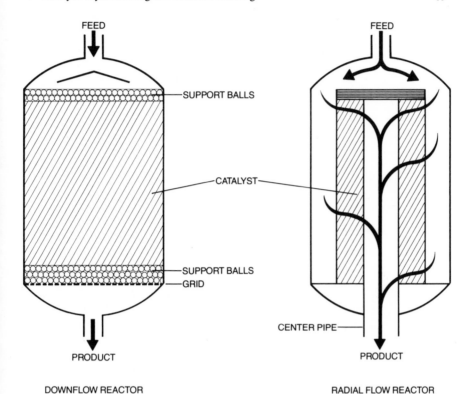

Fig. 12. Fixed bed reactor designs.

2.   Ebullating or Expanded Bed

This type of reactor is used in the H-Oil process (licensed by Hydrocarbon Research, Inc., and Texaco) and the LC-Fining process (licensed by The Lummus Company and Cities Service Research and Development Company) to hydrotreat and hydrocrack residual feedstocks. This type of reactor uses an internal recycle pump to provide an upward flow of liquid that expands the catalyst bed, resulting in a close approximation of a back-mixed isothermal reactor (see Fig. 13). The advantages of this expanded or ebullated bed reactor have been listed [5]:

(a)   Close control of the highly exothermic reactions,

(b)   A longer on-stream factor because the catalyst is added and withdrawn without shutting the reactor down,

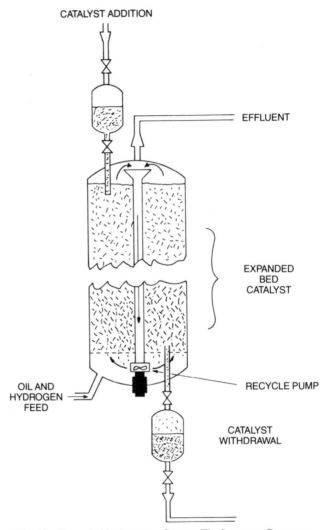

CATALYST ADDITION

EFFLUENT

EXPANDED
BED
CATALYST

OIL AND
HYDROGEN
FEED

RECYCLE PUMP

CATALYST
WITHDRAWAL

Fig. 13. Expanded bed reactor. Source: The Lummus Company.

(c)   maintenance of constant catalyst activity and, thus, constant product yields, product quality, and hydrogen consumption, and

(d)   elimination of catalyst bed plugging, channeling, and pressure drop buildup due to feed particulate contamination and/or coke buildup.

IV. Specific Process Applications

A. LIST OF APPLICATIONS BY FEEDSTOCK AND/OR PRODUCT FUNCTION

1. Lighter Than Naphtha (C_4 and Lighter)

HYDROGEN PLANT FEED. Essentially all hydrogen plant feed passes over a zinc oxide bed to remove H_2S and light mercaptans before going to the stream reforming catalyst. This normally is sufficient for feeds such as natural gas and LPG, but heavier feeds such as naphtha and reformer make gas (which contains some C_5 and heavier compounds) may require a hydrotreating catalyst in front of the zinc oxide bed.

CLAUS TAIL GAS. Some processes treating tail gas from Claus sulfur recovery plants, such as Beavon Sulfur Removal and SCOT, involve an initial hydrotreating step. Sulfur and SO_2 are hydrogenated and COS and CS_2 are hydrolyzed, reacting all sulfur to H_2S [6].

2. Naphtha (C_5 to 400°F)

LIGHT SR NAPHTHA (C_5 to 180°F). This naphtha is pretreated before being fed to an isomerization unit (or to a hydrogen plant, as discussed above).

HEAVY SR NAPHTHA (180–400°F). This is the primary feedstock for catalytic reforming, requiring essentially complete removal of sulfur and nitrogen (down to 1 ppm maximum each, and sometimes as little as 0.2 ppm maximum), as well as lead (to 10 ppb maximum) and arsenic (to 2 ppb maximum).

FLUID CATALYTICALLY CRACKED, COKER, AND VISBREAKER NAPHTHA. These cracked naphthas are relatively high octane but not too stable because of their high olefins content, including some diolefins. This is especially true of thermally cracked (coker and visbreaker) naphtha. These feeds frequently are blended with SR naphtha, to reduce exotherms and extend catalyst cycle life, before being hydrotreated.

PYROLYSIS NAPHTHA. This stream, sometimes called dripolene, is a by-product of olefins production (such as ethylene) by steam cracking. This naphtha is very high octane because it consists almost entirely of aromatics and olefins, but also is highly unstable because of a very high diolefin content (frequently 10 wt% or more). Most units hydrotreating pyrolysis naphtha are two-stage, with the first stage operated at a low temperature, frequently with a noble metals catalyst. This allows removal of the highly reactive diolefins before increasing the temperature in the second stage as required to remove sulfur and nitrogen.

3. Middle Distillate (350 – 650°F)

KEROSENE AND JET FUEL (ABOUT 325 – 525°F). The primary purpose is to reduce the aromatics content and thus to increase the smoke point. In this boiling range, sulfur normally is low enough that acceptable product levels are easily achieved by any hydrotreating conditions that will give the required smoke point. With a normal nickel – molybdenum or cobalt – molybdenum hydrotreating catalyst, high degrees of aromatics saturation are difficult to obtain at moderate pressure, partly because there are temperature equilibrium restrictions. As an example, see Fig. 14. More complete aromatics saturation can be provided by a number of proprietary processes, which frequently require two stages. The first stage is primarily standard hydrotreating to remove sulfur and nitrogen, and the second stage employs a noble metal catalyst to achieve much greater aromatics saturation at lower pressures and temperatures. These processes also can be used to produce special low-aromatics naphtha solvents.

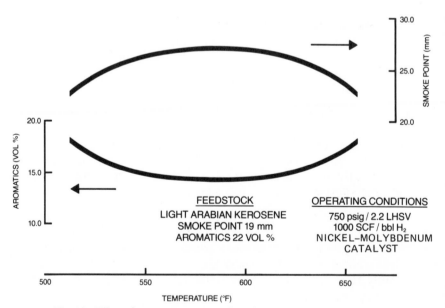

Fig. 14. Effect of temperature on smoke point and aromatics saturation.

DIESEL FUEL (ABOUT 400–600°F). The main objectives are sulfur reduction plus diesel index and cetane number improvement, which again is related to aromatics saturation.

NO. 2 FUEL OIL (ABOUT 400–650°F). This is the primary fuel oil for residential use, commonly called furnace oil or heating oil. The limiting factor for hydrotreating generally is product sulfur specification. Cracked stocks (from a coker, fluid catalytic cracker, or visbreaker) frequently are blended with SR for feed to this kind of hydrotreater.

4. Vacuum Gas Oil (650–1050°F)

LUBES, WAXES, AND WHITE OILS. In the United States, about 80% of finished lubricating oils receive some kind of hydrotreating [7]. About 70% is relatively low-pressure, low-temperature catalytic finishing. The other 30% is relatively high-pressure, high-temperature hydrogen processing, including low- to high-severity cracking. The product specifications of most concern are neutralization number, color, color stability, pour point, sulfur content, viscosity, and viscosity index. High-grade wax and white oil hydrotreating normally is done by special proprietary processes.

CAT FEED HYDROTREATING. Vacuum gas oil (VGO) comprises the bulk of feedstock going to a fluid catalytic cracker. Hydrotreating this feed can significantly increase conversion and improve product yields off the FCC, as well as solving SO_x and NO_x emissions problems. Catalytic feed hydrotreating benefits will be discussed further in Section IV.D.

HYDROCRACKER FEED PRETREATMENT. VGO also is the most common feed to a hydrocracker. Heavy cracked gas oils are fed to some units also. In any case, feed must be desulfurized and denitrogenated in a pretreatment stage to prevent poisoning of the downstream hydrocracking catalyst.

INDIRECT RESID DESULFURIZATION. As an alternative to the high-capital investment and operating costs associated with direct resid hydrotreating, some refiners desulfurize VGO and then blend it back into nonhydrotreated atmospheric or vacuum resid as required to meet no. 6 fuel oil sulfur specifications. However, since the weight percent sulfur level in the resid fraction is considerably higher than in the untreated VGO, the blended sulfur levels achievable by this procedure are not very low. With the current trends toward processing heavier (which means a higher ratio of resid to VGO) and more sour crude oils, it is becoming even harder to achieve acceptable sulfur levels this way.

5. *Residual Oil*

ATMOSPHERIC RESID (650°F+). Because of the reasons just discussed, atmospheric resid hydrotreating is becoming more widely used despite the high capital and operating costs involved. Only by resid hydrotreating can the high-molecular-weight asphaltenes and porphyrins be effectively attacked, thus greatly reducing Conradson carbon residue and contaminant metals, along with achieving maximum desulfurization. Direct resid hydrotreating therefore provides not only a good low-sulfur fuel oil but also an upgraded feedstock for further conversion downstream. By increasing operating severity, significant conversion by hydrocracking also can be achieved in the resid hydrotreater itself. Benefits due to this upgrading and higher conversions will be covered further in Section IV.D.

VACUUM RESID (1050°F+). Similar comments could be made about vacuum resid hydrotreating, except that it requires an even higher severity operation than atmospheric resid. At the present time, very little straight vacuum resid is being hydrotreated.

B. FEEDSTOCK PROPERTIES

The properties of four typical hydrotreating unit feedstocks are listed in Table II.

TABLE II

Properties of Four Typical Hydrotreater Feedstocks

Property	Naphtha, 85% SR/15% coker	Middle distillate, 60% SR/40% FCC LCO	VGO	Atmospheric resid
ASTM distillation (°F)	(D-86)	(D-86)	(D-1160)	(D-1160)
Initial boiling point	180	385	680	670
10%	235	440	800	780
50%	280	515	860	920
90%	340	570	990	—
End point	385	640	—	—
Gravity (°API)	54	31	20	15
Sulfur, (wt%)	0.10	1.4	2.4	3.3
Total nitrogen (wt%)	0.01	0.08	0.10	0.34
Olefins (vol%)	5	4	—	—
Pour point (°F)	—	—	95	130
Conradson carbon residue (wt%)	—	—	1.1	8.8
Nickel (wt ppm)	—	—	—	20
Vanadium (wt ppm)	—	—	—	70

C. OPERATING CONDITIONS AND RESULTS

Table III lists typical operating conditions and performance results for four types of units. For consistency, we have assumed that the feedstocks are the same four described in Table II. Further notes on this table are:

(1)   The performance results listed are intended as approximations, not exact predictions for the feedstock properties and operating conditions listed.

(2) H_2 consumption numbers are intended to represent total consumption, including chemical consumption, dissolved H_2, losses owing to leaks, etc. H_2 consumption will be discussed further under the subject of H_2 partial pressure in Section VI.

(3)   Start of run (SOR). Maximum end-of-run (EOR) temperatures vary widely from unit to unit depending on metallurgical restrictions and preheat furnace limitations, but 750°F maximum EOR is fairly typical for the reactor inlet temperature. Theoretically this temperature limit is what determines the cycle length, but in real life cycles frequently are ended for reasons other than catalyst deactivation, such as a pressure drop buildup or a scheduled maintenance shutdown. The cycle length figures listed are intended to be typical of actual commercial operating experience.

(4)   Total catalyst life will be discussed in Section V.

TABLE III

Typical Operating Conditions and Performance Results

Conditions or results	Naphtha, 85% SR/15% coker	Middle distillate, 60% SR/40% FCC LCO	VGO	Atmospheric resid
LHSV (hr^{-1})	4	3	1.5	0.5
Pressure (psig)	450	600	1000	1500
Treat gas rate (SCF/bbl)	600	800	1500	3000
H_2 purity (vol%)	75	80	85	85
H_2 consumption, as 100% H_2 (SCF/bbl)	75	175	250	700
SOR reactor temperature (°F)				
Inlet	585	580	650	660
ΔT	30	50	40	70
Outlet	615	630	690	730
Sulfur removal (%)	99.96[a]	90	85	85
Nitrogen removal (%)	99.5[a]	50	45	40
Cycle life (bbl/lb)	220	150	70	10

[a] Calculated from product of 0.5 ppm each sulfur and nitrogen.

D. DOWNSTREAM PROCESSING BENEFITS

In addition to the obvious benefits of being able to meet both finished product specifications (sulfur content, smoke point, neutralization number, etc.) and air quality regulations, hydrotreating provides many downstream processing benefits. Let us look at three examples.

1. Catalytic Reformer Pretreating

As discussed above, sulfur and nitrogen must each be reduced to 1 ppm or less in most units to achieve satisfactory reforming operations. Some bimetallic reforming catalysts require feed sulfur levels as low as 0.2 ppm to achieve optimum performance, especially C_5+ yields, H_2 make, and cycle life. Similarly, lead and arsenic must be removed to levels in the area of 10 ppb maximum lead and 2 ppb arsenic to maintain satisfactory total catalyst life.

2. Fluid Catalytic Cracker Feed Hydrotreating

Several studies have been done that demonstrate the benefits of FCC feed hydrotreating, particularly increased FCC conversion and improved product selectivity. One study [8] focused on the importance of PNA saturation in achieving these benefits. PNA saturation as a function of temperature is shown in Fig. 15. The feedstock used was a 50/50 blend of VGO and FCC cycle oil. Other operating conditions were 2 LHSV, 750 psig total pressure, and 2000 SCF/bbl H_2 treat gas rate. Note from Fig. 15 that, for this feedstock and these conditions, PNA saturation goes through a maximum equilibrium at about 700°F, similar to the effect observed for aromatics saturation of kerosene and jet fuel (see Fig. 14). At 700°F, sulfur removal was over 95% and total nitrogen removal about 55%. Figure 16 plots chemical H_2 consumption versus temperature, which, as expected, tracks PNA saturation very well. This H_2 uptake of a catalytic cracker feedstock is the best single measure of how much it has been improved. Finally, several microactivity test (MAT) runs were made to compare FCC performance with untreated versus hydrotreated feed. Results are presented in Figs. 17 and 18, plotted as a function of chemical H_2 consumption.

Another extensive study on the benefits of FCC feed hydrotreating was reported by Ritter et al. [9]. Depending upon the conditions used, absolute FCC conversion increases due to hydrotreating ranged from 8 to 23 vol%. Hydrotreating also significantly increased gasoline yields and decreased coke make, both on an absolute basis and relative to conversion.

Bridge et al. [10] also studied the effect of FCC feed hydrotreating on FCC operation, finding an increase of 5 to 17 vol% conversion at constant severity or 16 to 26 vol% conversion at constant coke make.

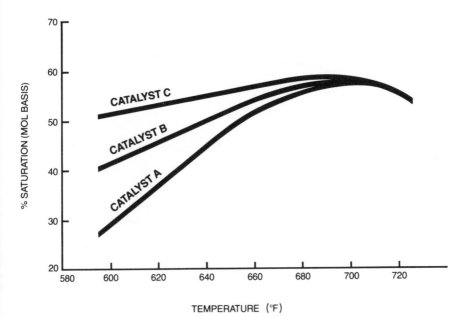

Fig. 15. PNA saturation versus temperature.

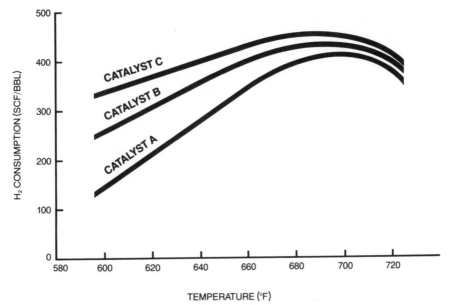

Fig. 16. Chemical H_2 consumption versus temperature.

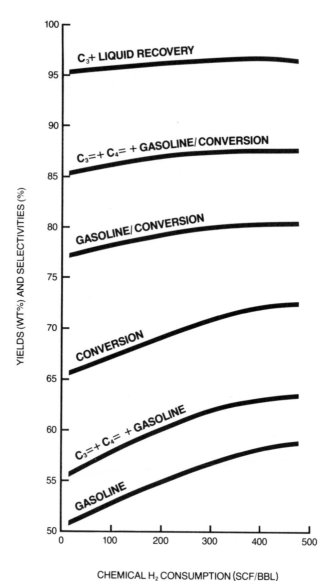

CHEMICAL H$_2$ CONSUMPTION (SCF/BBL)

Fig. 17. FCC yields versus H$_2$ consumption.

When it is necessary to reduce sulfur levels in FCC unit gasoline, hydro-treating of this gasoline would have the adverse effect of decreasing octane because of unavoidable olefin saturation. In such cases, therefore, FCC feed hydrotreating also offers the significant advantage of increasing gasoline octane by removing sulfur before the olefins are formed in the FCC unit.

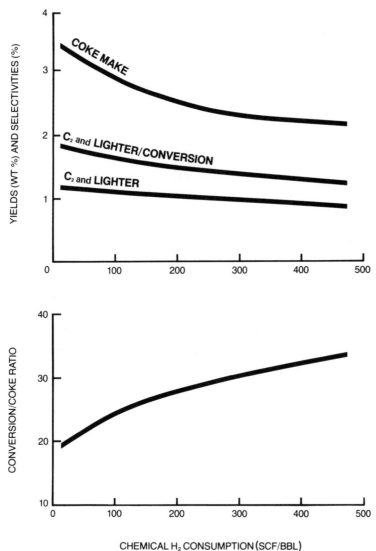

CHEMICAL H₂ CONSUMPTION (SCF/BBL)

Fig. 18. FCC yields versus H₂ consumption.

3. Residual Oil Conversion

As discussed previously, hydrotreating resids to provide desulfurization, demetallation, and reduction of Conradson carbon residue offers many downstream conversion benefits. Whittington *et al.* [11] tabulated these advantages:

(a) improves FCC feedstock quality (similar to what was discussed above for VGO and lighter feedstocks);

(b) reduces coke yield at delayed coker and increases liquid products production;

(c) increases coke selling price as a result of a lower sulfur and metals content;

(d) increases capability to process a wide variety of crudes;

(e) produces superior quality no. 2 fuel oil; and

(f) reduces the materials-of-construction cost for new FCC and coker processing as a result of low-sulfur feedstock.

This paper, plus an earlier one by Murphy and Treese [12], specifically discuss the benefits of Unicracking/HDS followed by the Kellogg Heavy Oil Cracking (HOC) process.

As mentioned before, significant conversion by hydrocracking can be achieved in a resid hydrotreater itself by increasing the operating severity. Ellis *et al.* [13] discuss a high-temperature operation strategy as an extension of Exxon's Residfining technology. Rather than the conventional gradual increase in temperature over a cycle, this strategy involves an accelerated progression whereby the temperature is increased rapidly from SOR to EOR conditions over a short interval, but done in such a way that the early catalyst deactivation rate is not excessive. The result is a rapid increase in the 1050°F+ conversion level from the 10–15% typical for SOR conditions up to the 45–50% normally not achieved until the EOR.

V. Catalysts

A. CHEMICAL AND PHYSICAL DESCRIPTION

Hydrotreating catalysts can be described as a mixture of transition metal compounds dispersed throughout a controlled surface area support. The support (or substrate) is primarily γ-alumina, sometimes mixed with silica. The metals normally are molybdenum (or sometimes tungsten) plus nickel and/or cobalt.

Either molybdenum or tungsten appears necessary to provide good activity. Cobalt and nickel do not provide significant activity when present alone, but they increase activity when combined with molybdenum or tungsten— hence cobalt and nickel are called promoter metals. Tungsten catalysts are typically promoted with nickel, and molybdenum catalysts with nickel or cobalt.

TABLE IV

Typical Hydrotreating Catalyst Properties

Chemical content and properties	A	B	C	D
Chemicals (wt% dry basis)				
MoO_3	15.0	18.5	16.2	13.5
CoO	3.2	—	2.5	3.2
NiO	—	3.3	2.5	—
SiO_2	—	—	—	4.0
Physical properties				
Surface area (m^2/g)	310	180	230	330
Pore volume (cm^3/g)	0.80	0.53	0.52	0.60
Diameter (in.)	0.125	0.062	0.050	0.038
Average length (in.)	0.23	0.18	0.16	0.13
Compacted bulk density (lb/ft^3)	36	52	46	44
Average crush strength/ length (lb/mm)	4.2	3.1	3.3	2.2

Table IV lists the chemical and physical properties of four typical hydrotreating catalysts, which can be described as:

(a) $\frac{1}{8}$-in. (3.2-mm) low-density cobalt–molybdenum catalyst,

(b) $\frac{1}{16}$-in. (1.6-mm) high-density nickel–molybdenum catalyst,

(c) $\frac{1}{16}$-in. (1.6-mm) equivalent diameter "shaped" cobalt–nickel–molybdenum catalyst,

(d) $\frac{1}{32}$-in. (0.8-mm) nominal diameter cobalt–molybdenum catalyst, containing some silica, designed for ebullating or expanded bed units.

With the exception of normal contaminants such as SO_4 (typically 0.3–1.0 wt%), Na_2O (0.03–0.09 wt%), and Fe (0.03–0.05 wt%), the balance of the chemical composition in Table IV is alumina. The four catalysts listed are all extrudates, which is the most common form used today. Some catalysts are also available in tablet and spherical form.

Nickel–tungsten catalysts are more expensive than Ni–Mo or Co–Mo catalysts and are used primarily for special applications where feed sulfur content is reasonably low and very high saturation and/or moderate cracking activity is required. Metals levels are high, typically 5–20% NiO and 20–25% WO_3, frequently on a silica–alumina base.

Another important characteristic of hydrotreating catalysts, especially when treating heavier oils, is pore size distribution. This is measured by a mercury porosimeter, which applies high pressures to force mercury into the pores of a catalyst sample. For a given pressure applied, mercury penetrates

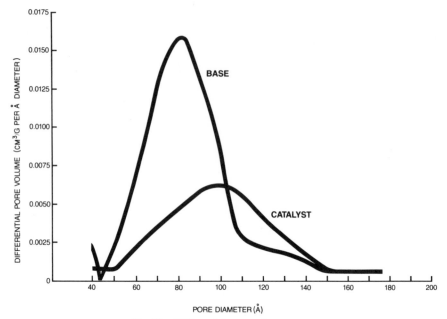

Fig. 19. Differential pore volume curves.

into pores of a specific pore diameter (the higher the pressure, the smaller the diameters penetrated).

Figure 19 shows differential pore size distribution curves for a typical finished hydrotreating catalyst and a corresponding base (or substrate) support before addition of the active metals. The peak of each curve is called the pore mode, for these examples about 83 Å diameter for the base and 99 Å for the catalyst (this apparent shift in pore mode is actually due to the contact angle of mercury being different before and after metals addition). Some catalysts and bases, called bimodal, have two distinct peaks in their differential pore volume curves.

A quick method sometimes used to estimate average pore diameter without measuring the pore size distribution is to divide the surface area (m^2/g) into four times the pore volume (cm^3/g) times 10^4. As an example, for catalyst C in Table IV, $4 \times 0.52 \times 10^4 \div 230 = 90$ Å average pore diameter.

In recent years, the use of shaped catalysts has increased greatly because of the benefits offered by their higher ratio of external (or geometric) surface area per unit volume. Richardson *et al.* [14] list these benefits: (1) high crush strength, (2) higher contaminant metals tolerance, (3) higher diffusion rates, (4) lower pressure drop buildup rate.

B. CATALYST SELECTION

1. Relative Activity Definition

When running pilot plant tests to compare the performance of various catalysts on a given feedstock and for a set of operating conditions, a consistent means is required for determining the relative activity of each catalyst.

To avoid any estimations of reaction order, some refiners compare catalysts by varying the space velocity as necessary to obtain a given conversion level, with all other conditions (pressure, temperature, and H_2 treat gas rate) held constant. Then, the relative activity (RA) of catalyst A versus that of catalyst B is calculated simply as

$$RA_{A\,vs\,B} = (SV)_A/(SV)_B.$$

This procedure, however, has the disadvantage of requiring several data points (several different space velocities) for each catalyst tested. Therefore, most refiners run each catalyst under the same conditions and use an estimated reaction order to calculate relative activity. The reaction orders most commonly applied are 1, 1.5, and 2. Integrating the basic rate equations gives

$$\text{for 1st order:}\quad RA_{A\,vs\,B} = \frac{\ln(100/100 - C)_A}{\ln(100/100 - C)_B}$$

$$\text{for 1.5 order:}\quad RA_{A\,vs\,B} = \frac{(\sqrt{100/100 - C} - 1)_A}{(\sqrt{100/100 - C} - 1)_B}$$

$$\text{for 2nd order:}\quad RA_{A\,vs\,B} = \frac{(C/100 - C)_A}{(C/100 - C)_B},$$

where C is the percent removal of sulfur (or nitrogen, metals, etc.).

These calculations provide either a relative weight activity (RWA) or relative volume activity (RVA), depending upon whether catalysts are compared on an equal weight or volume basis.

2. Choice of Nickel–Molybdenum versus Cobalt–Molybdenum Catalyst

In general, cobalt–molybdenum catalysts are preferred for desulfurization of SR feedstocks, although nickel–molybdenum and cobalt–nickel–molybdenum catalysts can given equally good sulfur removal results for certain feedstocks and operating conditions. On feedstocks containing high nitrogen levels and/or cracked feedstocks, nickel–molybdenum catalysts may be better even for sulfur removal.

Nickel–molybdenum catalysts generally are better for nitrogen removal and for aromatics saturation. In applications where both sulfur removal and nitrogen removal (or aromatics saturation) are critical, the choice of Ni–Mo versus Co–Mo may depend on which requirement is the more difficult to achieve.

Ni–Mo catalysts generally are more responsive to differences in H_2 partial pressure so that, everything else being equal, higher-pressure operations favor the performance of Ni–Mo versus Co–Mo catalysts. This is not to say that a Ni–Mo catalyst always performs better than Co–Mo at high pressures, only that Ni–Mo performs *relatively* better at high pressures than at low pressures.

3. Pilot Plant Evaluations

Most major refiners base their catalyst selections largely upon pilot plant comparison of the various catalysts available. The type of pilot plant units most commonly used are small trickle bed downflow reactors, with frequently only about a 1-in. inside diameter and containing 50–150 cm^3 of catalyst often diluted with small glass beads or sand.

Two factors that make absolute results different in a pilot plant unit and a commercial unit are

(a)   Thermal mode of operation—because of the relatively large reactor and furnace heat sink versus reaction exotherm, pilot plant reactors are more isothermal than commercial reactors (which are close to adiabatic).

(b)   Mass flow velocity—small pilot plants are typically run with mass flow velocities between 25 and 100 lb/hr ft^2, whereas most commercial units operate in the range 1000–3000 lb/hr ft^2. This difference is the main reason for diluting the catalyst bed in pilot plant testing, which helps to reduce wall effects, increase liquid holdup, and minimize channeling.

Therefore considerable caution should be used when applying pilot plant data to a commercial operation. However, *relative* differences observed as a result of changing catalysts should be valid so long as identical pilot plant test units, feedstock, and operating conditions are used to compare all catalysts.

4. Commercial Comparisons

Although optimum performance on a commercial unit is always the desired end result, commercial comparisons of catalyst performance frequently are not very meaningful. This is because it is nearly impossible to maintain constant feedstock and operating conditions on a given unit from one cycle to the next or to adjust for all the differences that do occur (in many cases, these differences are not even recognized).

C. CATALYST HANDLING

1. Containers

The most common container for shipping fresh hydrotreating catalyst is a 55-gal steel drum (frequently reconditioned), typically holding about 300 lb of catalyst. Also frequently used are reinforced paper or plastic bags that typically hold about 1200 lb of catalyst.

The general advantages of steel drums are greater resistance to crushing during handling, withstanding prolonged outside storage better, easier reuse for storing or shipping spent catalyst (or other materials), and higher resale value.

The general advantages of reinforced paper or plastic bags are less need to use pallets, easy unloading through bottom opening, lower tare transportation cost, fewer containers per shipment, and easier disposal of empty containers (or less storage space if kept for reuse).

2. Reactor Loading

BED SUPPORTS. For reasons discussed in Section III.D, downflow fixed bed reactors use support balls above and below the catalyst bed. NPRA [15] discusses the loading and size distribution of these balls. A typical loading pattern from top to bottom is 6 in. of $\frac{3}{4}$-in. balls, 6 in of $\frac{1}{4}$-in. balls, catalyst bed ($\frac{1}{8}$ or $\frac{1}{16}$ in.), 6 in. of $\frac{1}{4}$-in. balls, 6 in. of $\frac{3}{4}$-in. balls, screen support grid.

When a bottom support grid is not used, then $\frac{3}{4}$-in. balls typically fill the bottom dished head plus 4–6 in. above the screened (or slotted) outlet pipe entrance. It is important that each layer of support balls be reasonably level over the entire reactor cross section. If multiple catalyst beds are used in one reactor, a support screen and graded layers of support balls are required for each bed.

CATALYST LOADING METHODS. The traditional "hopper-and-sock" method of loading catalyst uses a flexible hose that can be shortened as the reactor fills up so that catalyst free-fall distance is maintained at between 1 and 4 ft. This method minimizes catalyst breakage but generally results in a relatively low reactor loaded density, typically only 85–90% of the compacted bulk density (CBD) value measured in the laboratory. In order to increase the amount of catalyst that can be loaded into existing commercial reactors (or to decrease the size of new reactors), many refiners use various dense loading techniques. Snow and Grosboll [16] discuss the catalyst-oriented packing (COP) method developed by Atlantic Richfield Company, which has increased loaded densities by 10–15% over conventional methods and significantly improved catalyst–liquid contact efficiency in some units.

FEED FILTRATION AND FLOW DISTRIBUTION. All downflow mixed phase reactors have some kind of flow distributor to split inlet feed uniformly over the cross-sectional area (and to redistribute feed between catalyst beds of a multibed reactor). As discussed above, support balls also help to distribute feed, as well as to reduce the bed plugging effects of particulate matter entering with the feed. In addition, many units screen out particulate matter by the use of trash baskets, which are wire mesh cylindrical baskets inserted 1–3 ft down into the catalyst bed, so that the cross-sectional area for feed entering the catalyst is considerably increased. Some units also use guard reactors and/or inlet feed filters to maximize the removal of particulates.

3. Reactor Unloading

Most refiners have their own specific procedures for unloading catalysts, with a great deal of attention paid to potential toxicity problems. Many refiners use outside companies specializing in catalyst handling. In general terms only, the steps commonly involved in catalyst unloading are:

CATALYST STRIPPING. If a catalyst is being dumped unregenerated, it should be stripped with hydrogen recycle gas (after the feed is cut out) to maximize the removal of hydrocarbons. For resid and heavy gas oil operations, it may be desirable to strip the catalyst with a light liquid feed prior to hydrogen stripping.

COOLING. Regenerated catalyst is normally cooled down with once-through steam (assuming steam–air regeneration) to 400 to 500°F and then switched to nitrogen or air for cooling the rest of the way down to 100 to 200°F. Unregenerated catalyst is normally cooled down with circulating hydrogen and then purged thoroughly with nitrogen before the reactor is opened. Below 300 to 400°F, the formation of highly toxic carbonyls (especially nickel carbonyl) is of great concern, so that gases used for cooling must be CO-free.

CATALYST DUMPING AND REACTOR ENTRY. Unregenerated catalyst should be dumped under an inert atmosphere after being cooled below 200°F. If reactor entry is required, most refiners require cooling to 100 to 120°F. Whether catalyst has been regenerated or not, any reactor entry work should be done only with fresh-air breathing equipment.

SCREENING. Screening is required to separate the catalyst and support balls into their proper sizes.

D. PRESULFIDING

1. Purpose of Presulfiding

Hydrotreating catalysts are manufactured and sold with metals (cobalt, nickel, molybdenum, etc.) in an oxide state. These metals must be converted to the proper sulfided state in order to achieve the desired activity and selectivity.

Because metals are returned to their oxide form during regeneration, presulfiding is required for regenerated catalysts also.

2. Typical Procedures

Most commonly recommended commercial presulfiding procedures include the use of a liquid SR feedstock spiked with an easily reacted sulfiding agent such as carbon disulfide, dimethyl sulfide (DMS), dimethyl disulfide (DMDS), or light mercaptans. These sulfiding agents allow a complete reaction to the active metal forms at relatively low temperatures, thus minimizing premature coking and any reduction of metal oxides to the unreactive metallic state. Cracked feedstocks should not be used until the catalyst is fully sulfided, again to avoid premature coking.

Table V gives a brief outline of a typical presulfiding procedure. The

TABLE V

Typical Commercial Presulfiding Procedure

Step	Procedure
1	After thorough purging with inert gas, pressure to 200–500 psig with dry hydrogen, establish the maximum once-through rate, heat to 300–350°F at 100°F/hr, and hold until no more water is condensing out in the separator
2	Establish the normal recycle or once-through treat gas rate and introduce SR feedstock spiked with sulfiding agent equal to 1–2 wt% sulfur
3	Increase the temperature as required (to achieve good conversion to H_2S), but not above 500°F anywhere in the catalyst bed, and continue sulfiding until a copious breakthrough of H_2S is observed
4	Stop the sulfur injection, wait 1 hr, and check again for a continuing H_2S breakthrough. Resume sulfur injection as necessary until the H_2S breakthrough holds at a high level
5	Increase the temperature at 25°F/hr to 550–600°F and continue sulfiding until a second breakthrough of H_2S is observed
6	Hold for 30 min to ensure complete sulfiding. To fully sulfide most catalysts, about 0.11 lb of spiking agent sulfur per pound of catalyst is required
7	Remove excess H_2S from the recycle gas by placing a scrubber in service (if available) or purge with fresh hydrogen
8	Adjust the reactor temperature, pressure, and feed rate to normal (or anticipated) SOR conditions

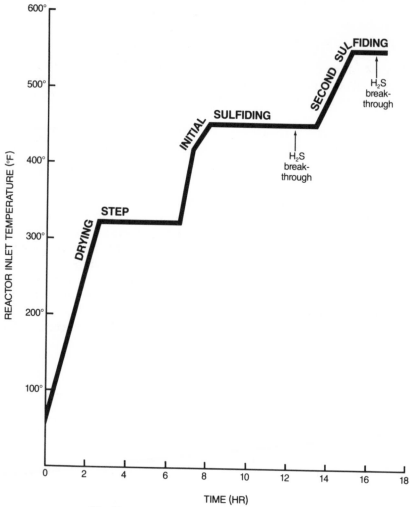

Fig. 20. Time–temperature profile for presulfiding

approximate time–temperature profile for such a procedure is shown in Fig. 20.

An alternative presulfiding procedure using H_2S in H_2, without any liquid feed present, is similar to the standard procedure used by many refiners and catalyst suppliers for pilot plant testing. Such a procedure can be used successfully in commercial units, but the spiked liquid feed procedure normally is preferred to achieve better temperature control (because of the greater heat sink) and good catalyst wetting.

E. DEACTIVATION

Carberry [17] divides catalyst deactivation into three categories:

(1) Poisoning—chemisorption of reactants, products, or impurities that occupy sites otherwise available for catalysis,

(2) Fouling—physical blockage of catalyst surface and pores by surface reactions (such as some coking reactions) or deposition of particulate matter in the feed, and

(3) Sintering or phase transformations—significant alteration of the catalyst's physical structure and/or chemical nature.

For commercial hydrotreating operations, the items normally of most concern are coking, metals and other inorganic poisons, and physical degradation.

1. Coking

The most common cause of hydrotreating catalyst deactivation is coking. As coke builds up, operating temperatures must be increased to maintain the desired unit conversions. At some point, the unit furnace capacity or metallurgical limits (or excessive yield loss due to hydrocracking) prevent further temperature increases. Unless the refiner is willing to decrease the unit throughput, the end of the cycle has been reached.

Therefore, the estimation of coke deactivation rates is a major part of the hydrotreating unit design. Directionally, coking rates increase with increasing temperature, decreasing hydrogen partial pressure, increasing conversion levels (such as percentage of sulfur removal), increasing feed boiling range (especially feed end point), and increasing percentage of cracked stocks in the feed. New unit space velocities are selected such that the required conversion levels can be achieved at a reasonable temperature and hydrogen partial pressure, thus resulting in a reasonable cycle life. Of course, what is considered reasonable varies greatly depending upon both feedstock properties and product requirements. Coke levels on the catalyst at the end of the cycle can vary from only 3–4 wt% for a SR light naphtha service to over 25 wt% for a residual oil hydrotreater.

Figure 21 shows a typical time–temperature deactivation curve for a hydrotreating catalyst, which can be broken down into three sections:

(a) Initial stabilization—typically an increase of 10 to 20°F over the first few days of operation,

(b) Regular deactivation—more or less linear deactivation rate, and

(c) Accelerating deactivation—rapidly increasing deactivation rate

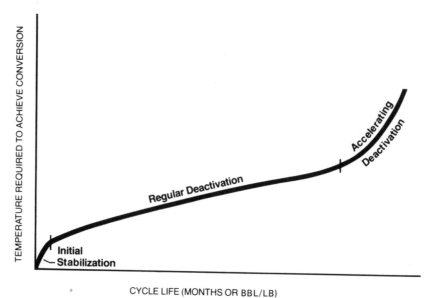

CYCLE LIFE (MONTHS OR BBL/LB)

Fig. 21. Time–temperature deactivation curve.

due to the snowballing effects of coking and increasing temperature; heavy-metals buildup, such as during resid hydrotreating, also can contribute greatly to this accelerating deactivation period.

2. Metals and Other Inorganic Poisons

NICKEL AND VANADIUM. These are the two most prevalent metals in crude oil. They are primarily contained in porphyrin-type compounds (as discussed in Section II) and are concentrated in the residual oil boiling ranges. Although Ni and V are hydrotreating catalyst poisons, these compounds must be attacked in high-metals feedstocks in order to achieve required high desulfurization levels and/or to pretreat feed properly for downstream conversion operations. Numerous articles have been published recently on this subject [10, 12, 18, 19]. The total weight of Ni and V in atmospheric resid feedstocks can vary from as low as 5 ppm to over 200 ppm. Concentrations in vacuum resid feedstocks are roughly twice as high. Total amounts of Ni and V deposited on spent hydrotreating catalysts vary greatly, depending upon feedstock metals levels, catalyst type, and unit performance requirements. Typical levels are between 10 and 30% by weight of spent catalyst. Some catalysts specifically designed for demetallation end up with 50% metals or 100% fresh catalyst weight. Shaped catalysts are becoming increasingly popular for metals removal because their greater ratio of external (geometric) surface area to volume allows a higher metals accumulation for a given amount of activity loss. In resid hydrotreating units containing two or more reactors in series, it is possible to follow the progress of metals buildup during a cycle by observing shifts in reactor exotherms. Shuke [20] explains how this can be done and the benefits resulting from it.

LEAD AND ARSENIC.   These two metals are present at much lower levels than nickel or vanadium in crude oils. Nevertheless, their levels in the gasoline boiling range frequently are high enough to cause problems, primarily because they are such severe poisons for reforming catalysts. As discussed above, naphtha reformer pretreaters must remove these metals down to levels of 10 ppb maximum for lead and 2 ppb maximum for arsenic. Lead and arsenic poison hydrotreating catalysts also, but normally breakthrough of lead or arsenic to the downstream reformer becomes a problem long before the loss of hydrotreating catalyst activity becomes significant. As a rule, arsenic tends to "chromatograph" better than lead from the top to the bottom of a downflow reactor bed, which may be one reason why sudden unexpected breakthroughs are more common with lead than with arsenic. Periodic high levels of lead in the feed (such as could occur from processing leaded gasoline slops) also can temporarily overload the hydrotreating catalyst metals removal capacity and cause a lead breakthrough. For this reason, some refiners limit allowable lead in the pretreater feed to about 100 ppb maximum. The absolute amount of lead and arsenic a hydrotreating catalyst can hold before breakthrough occurs is a function of feed concentration, catalyst type, unit operating conditions, and the geometry of the catalyst bed. Typical average levels for spent naphtha hydrotreating catalysts (not necessarily changed out because of breakthrough) are 0.05–0.5 wt% lead and 0.02–0.2 wt% arsenic.

IRON.   Iron enters hydrotreating reactors primarily as particulate scale and as such causes problems primarily because of bed plugging and secondarily because of catalyst pore plugging. Typical iron levels on spent catalysts are 0.1–1.0 wt%.

SODIUM.   Similar to iron scale, sodium salts entering hydrotreating reactors can cause bed plugging and catalyst pore blockage. In addition, sodium tends to promote catalyst sintering during regeneration. Levels in excess of 1% have been reported to cause severe activity declines. Typical sodium levels on spent catalyst are 0.1–0.5 wt% as Na_2O.

SILICON. Most silicon appears to come from antifoam agents used various places upstream in a refinery, most frequently in cokers or visbreakers. Again, bed plugging and pore blockage are the major problems caused. Levels as high as 30 wt% SiO_2 have been reported on spent catalyst; at which point activity is greatly reduced. More typical levels are 1–2 wt%, which do not appear to affect activity significantly.

SULFATES. High sulfate levels contribute to lower crush strength and lower activity, especially after regeneration, perhaps because of the formation of aluminum sulfate by reaction with the catalyst base. However, these effects are not well quantified. Typical levels on spent catalyst are 1–5 wt% as SO_4.

CHLORIDES. If entering chlorides are inorganic (most likely NaCl), they will deposit out on the catalyst as described above for sodium. Resultant chloride levels normally encountered are not high enough to affect the hydrotreating catalyst significantly. If entering chlorides are organic (most likely from chlorinated solvents used in oil production), then HCl will be formed, posing a considerable threat to downstream equipment corrosion and blockage (such as by reaction with NH_3 to form NH_4Cl deposits). After some very unpleasant experiences, many refiners now watch closely for excessive chloride levels in their crude oil.

CO AND CO_2. Based on comments from several sources, it appears that CO and CO_2 levels in H_2 treat gas should be held below 2 vol% total for good desulfurization results and below 0.5 vol% for good denitrogenation results. As discussed before, any gas used for cooling down reactors in preparation for opening must be CO-free in order to minimize the dangers of nickel or cobalt carbonyl formation.

3. Physical Degradation

Most hydrotreating catalysts in commercial use today are reasonably rugged and resistant to physical damage so long as proper catalyst handling procedures are used. The potential for permanent physical damage is greatest during regeneration of the catalyst, when excessive temperatures (especially localized hot spots) can be reached. As just discussed, certain poisons such as sodium and sulfate can promote catalyst sintering (loss of surface area), reduced crush strength, etc. Proper regeneration procedures will be discussed below.

4. Catalyst Life

Hydrotreating catalyst cycle life and total life, of course, vary widely depending upon feedstock properties and unit operating conditions. It also should be obvious from the above discussions about catalyst coking, poisons, regeneration, etc., that the estimation of actual catalyst life is a rather inexact science.

Nevertheless, we can mention some approximate life figures. Table III listed cycle lives for four typical hydrotreating operations. Expanding this information to include normal ranges gives the accompanying tabulation. It should be noted that commercial catalyst cycle life and total life are determined not only by deactivation but also frequently by other considerations such as the convenient time to shut a unit down and the need to ensure a minimum 1- or 2-yr run length the next time.

	Commercial catalyst life (bbl/lb)	
Feedstock boiling range	Cycle	Total
Naphtha	125–600	300–1200
Middle distillate	75–250	150–600
VGO	40–100	40–200
Atmospheric resid	5–20	5–20

F.   REGENERATION

1.   Purpose and Limitations

If catalyst deactivation is caused primarily by coking, most of this activity loss can be recovered by burning the catalyst in a controlled atmosphere of steam and air or nitrogen and air. However, if a significant amount of deactivation is due to poisoning by metals or other inorganic contaminants, even the most closely controlled and thorough regeneration will still leave the catalyst well below fresh catalyst activity.

Similarly, if pressure drop problems develop during a run because of the formation of polymers and gums (Section VI.F), a careful regeneration can burn them off. However, if the pressure drop buildup is a result of scale or other inorganic particulate carryover to the catalyst bed, then catalyst dumping and screening (with or without regeneration) generally will be necessary.

2.   Temperature Limits

Excessive temperature during regeneration is the most common cause of permanent catalyst damage (and sometimes of permanent reactor damage). A few examples are

(a)   Appreciable loss of catalyst crush strength can occur above 1200°F. The formation of inactive α-alumina above 1500°F decreases both crush strength and activity drastically.

(b)   Molybdenum sublimation can occur above 1300°F, which is evidenced by needlelike silvery recrystallization of MoO_3 on contact with cooler spots in the catalyst bed, on the reactor walls, or on downstream piping.

(c)   Catalytically inactive complexes can be formed with cobalt and alumina ($CoAl_2O_4$) above 1200°F and with nickel and alumina ($NiAl_2O_4$) above 1000°F.

To avoid such problems, it is also necessary to provide a safety factor allowing for both undetected hot spots and the differences between catalyst surface temperatures and bulk temperatures. Therefore, it is common to recommend limiting catalyst bed temperatures to 850°F maximum during regeneration.

3.   In Situ *versus Off-site Regeneration*

This subject will be covered in Chapter 2, Volume 3. Briefly, both *in situ* (catalyst remaining in the reactor) and off-site regenerations are widely used. Off-site regeneration requires a spare charge of catalyst on hand (or an extra long unit downtime), but this procedure can offer several advantages, such as

(a)   Improved contacting of catalyst and regeneration gases, resulting in both better temperature control and more complete regeneration,

(b)   Eliminating any air pollution problems due to regeneration effluent gases, and

(c)   Removal of fines by screening the catalyst after regeneration.

4.   *Typical* in situ *Procedure*

Table VI briefly outlines a typical *in situ* regeneration procedure for commercial units.

5.   *Activity Recovery and Significance*

If no significant physical damage or inorganic poisoning has occurred, a good regeneration can restore hydrotreating catalysts to essentially 100% of fresh catalyst activity, although more typical results fall between 75 and 95%. To translate these numbers into required unit operating temperatures, an activity recovery of, say, 80% would require about 10–15°F higher SOR

TABLE VI

Typical Commercial *In Situ* Regeneration Procedure (Steam – Air)

Step	Procedure
1	Cut the oil feed and purge with H_2-rich treat gas to strip out residual oil remaining on the catalyst
2	Connect the steam supply to a preheat furnace inlet, the air supply to reactor inlet, and the reactor outlet to effluent gas cleaning–disposal system
3	Cut the H_2-rich treat gas flow and introduce steam at a minimum mass flow velocity of 400 lb/hr ft², holding the reactor outlet pressure below 100 psig
4	Start firing furnace to increase the temperature to 700°F uniformly throughout the catalyst bed
5	Start the airflow at about 0.5 vol% oxygen at the reactor inlet and adjust it (up to 1.5 vol% O_2 maximum) to maintain a maximum flame front temperature of 750°F
6	After the initial flame front has passed through the reactor at 750°F, carefully increase the air feed, holding the maximum temperatures anywhere in the bed at 850°F. Typically not more than 2.0 vol% O_2 should be needed. After this initial burn is completed, continue the flow until at least 0.25 lb air per pound of catalyst has passed through the bed
7	Stop the airflow and continue to purge the reactor with steam for 30 min. Reduce the furnace firing until the bed temperatures reach 400–500°F, cut the furnace fires, and stop the steam flow. The system can be further cooled with nitrogen or air if catalyst dumping and/or reactor entry is required

versus the fresh catalyst. The effect on cycle life depends primarily on how much excess furnace capacity is available, thus determining the allowable temperature increase from the start to the end of the cycle.

Each refiner must assess his own economics for each unit to determine how many times he can afford to regenerate a batch of catalyst. Some units are so tight that not even an 80 or 90% activity recovery is acceptable, and these units use only fresh catalyst. Other units have such mild requirements that only 50% of fresh activity is still good enough, so multiple regenerations may be practical. Some refiners "cascade" catalyst—that is, they use new catalyst in their most critical units and then move it to less demanding units after regeneration.

As mentioned previously, it normally is not practical to regenerate hydro-treating catalysts containing appreciable amounts of contaminant metals. Residual oil catalysts, which are highly contaminated with Ni and V, are not regenerated. VGO catalysts may or may not be regenerated, depending upon the metals buildup and the physical condition of the catalyst. Most middle-distillate and lighter catalysts are regenerated at least once.

G. METAL RECLAMATION

At some point in the life of every hydrotreating catalyst charge, for the reasons discussed above, the refiner decides that it is no longer economical to continue using it. At one time, most refiners simply discarded this spent catalyst as landfill, but this disposition is rarely acceptable anymore, either environmentally or economically. Therefore, most spent hydrotreating catalysts today are sold to companies specializing in the recovery of one or more of the metals contained on the catalyst.

H. TOXICITY

The health and safety hazards associated with the handling of hydrotreating catalysts can be attributed primarily to the three most commonly contained transition metals: nickel, cobalt, and molybdenum. In the United States, the Occupational Safety and Health Act (OSHA) standards shown in the accompanying tabulation have been set for maximum allowable atmospheric contamination based on an 8-hr time-weighted average exposure. To ensure compliance with these standards, good ventilation and appropriate respiratory and other protective equipment must be provided.

Contaminant	mg/m^3
Cobalt (metal fume and dust)	0.1
Nickel (metal and soluble compounds, as Ni)	1
Molybdenum	
Soluble compounds	5
Insoluble compounds	15
Nuisance dust	15

Other specific toxicity concerns during and after the use of these catalysts involve nickel carbonyl (mentioned in Section V.C) and nickel subsulfide, which has been identified as a carcinogenic compound.

Catalyst suppliers can provide more detailed information on these hazards plus recommended safe handling procedures.

VI. Operating Performance Evaluation and Troubleshooting Guidelines

A. NORMALIZING DATA

Probably the most common sin of omission when evaluating commercial operating performance is a failure to normalize the data properly. This is the main reason why it is very difficult to compare the hydrotreating catalyst performance of two different units, or even of two operating cycles on the same unit. Frequently, a refiner is not aware of the importance of seemingly small differences in feedstock properties or operating conditions. Even if he is aware, frequently at least one or two pieces of required comparative data are unavailable, incomplete, or erroneous.

Therefore, a refiner must be very careful and thorough to be sure that all relevant operating data are accurately measured, correctly recorded, and properly analyzed for their effect on unit performance.

B. HYDROGEN PARTIAL PRESSURE

This subject was covered in considerable detail in an earlier paper [21]. As noted, hydrogen partial pressure is an often misunderstood variable. Improper calculations, mainly resulting from not taking all the necessary factors into account, are common, and many refiners focus too much on conditions at the reactor inlet rather than at the outlet. Inlet conditions are better known, easier to calculate, and less variable, but outlet conditions generally have a greater effect on catalyst life and unit performance. Some operations, such as all vapor phase SR naphtha hydrotreating, show little difference between inlet and outlet conditions, but this difference can be very significant in mixed phase high H_2 consumption hydrotreaters.

1. Variables to Consider

Variables affecting H_2 partial pressure at the reactor inlet are (a) total pressure, (b) amount of feed varporization, (c) amount of dissolved gases (for mixed phase operations), and (d) treat gas composition.

Feed vaporization and dissolved gases must be determined by vapor–liquid equilibrium calculations.

Variables affecting H_2 partial pressure at the reactor outlet include all of the above plus the pressure drop (to determine the outlet pressure) and chemical H_2 consumption.

2. Hydrogen Consumption

Chemical H_2 consumption includes H_2 required for (a) impurities removal (primarily the heteroatoms sulfur, nitrogen, and oxygen), (b) olefin and diolefin saturation, (c) aromatics saturation, and (d) hydrocracking (generally low under normal hydrotreating conditions).

For impurities removal, hydrogen is required not only to remove the heteroatom (as H_2S, NH_3, or H_2O) but also to replace the heteroatom in the hydrocarbon molecule left behind (refer to Figs. 2–4). Therefore, the amount of H_2 required varies considerably depending upon the type of sulfur, nitrogen, and oxygen compounds being reacted. As a rough rule of thumb, typical H_2 consumption values can be estimated as 100 SCF/bbl per each weight percent sulfur removed, 300 SCF/bbl per weight percent nitrogen, and 250 SCF/bbl per weight percent oxygen (the oxygen content generally is small, except for feeds derived from synthetic fuels such as coal).

The amount of net makeup gas required depends on both the H_2 concentration of the gas and the amount of *total* H_2 consumption, which includes not only chemical consumption but also an allowance for leakage losses (about 10 SCF/bbl is typical) and dissolved H_2. As it contributes to the total hydrogen consumption (rather than to reactor H_2 partial pressure), dissolved H_2 must be determined under conditions existing in the high-pressure separator because this is where liquid product is separated from the recycle (or once-through) gas stream. Depending on liquid product properties and separator operating conditions, dissolved H_2 typically varies between 10 and 30 SCF/bbl.

3. General Observations and Conclusions

By applying the preceding approach and calculating H_2 partial pressures for a variety of feedstock properties and operating conditions, the following general observations and conclusions can be made:

(a)   In units where the amount of hydrogen consumption is considerable, attention should be focused on reactor outlet rather than inlet H_2 partial pressure.

(b)   When feedstocks remain at least partly liquid under reaction conditions, the H_2 partial pressure can decrease significantly from the SOR to the EOR, largely because of more feed vaporization as temperatures are increased to compensate for catalyst deactivation. Note that this lower H_2 partial pressure combines with the higher temperatures themselves to have a snowballing effect on coke laydown, thus tending to increase catalyst deactivation rate as the run progresses.

(c)   The importance of changes in treat gas rate is highly dependent upon the ratio of SCF/bbl chemical H_2 consumption to SCF/bbl H_2 in the treat gas. When treat gas rates and/or hydrogen purity causes this ratio to approach unity, the reactor outlet H_2 partial pressure falls off rapidly.

(d) In applications where the treat gas rate is much greater than the hydrogen consumption, the H_2 partial pressure cannot be increased much further without increasing the total system pressure.

C. HYDROGEN TREAT GAS RATE AND RESIDENCE TIME

Primarily because of rapidly rising energy costs over the last decade, many refiners have considered reducing gas compression requirements for hydrotreating units, either by decreasing treat gas rates or by using hydrogen from low-pressure sources without the aid of booster compressors.

A typical recommendation is that the H_2 contained in the total treat gas entering the reactor be at least three times the chemical H_2 consumption. Lower treat gas rates obviously can be used, but significant penalties in higher unit operating temperatures and a lower catalyst cycle life will be seen. Conversely, considerably higher treat gas rates (perhaps five to six times the chemical consumption) can be justifiable in some units.

It should be noted that it is possible to have too *much* treat gas for technical, as well as economic, reasons. Edgar [22] reported on a pilot plant gas oil hydrotreating study that varied treat gas rates over a wide range. Some of these results are plotted in Fig. 22. The decrease in sulfur removal for the 650°F curve at H_2 rates above 1400 SCF/bbl is believed to be real. At such relatively high treat gas rates (over six times the chemical H_2 consumption in this case), additional incremental hydrogen does not contribute significantly to increasing the hydrogen partial pressure. Instead, the higher treat gas rates appear to be detrimental because they reduce liquid holdup in the catalyst bed, thus reducing the residence time of the oil in contact with the catalyst.

D. REACTION INHIBITION BY H_2S

It is well known (and kinetically logical) that H_2S has a reversible inhibitory effect on desulfurization reactions, but the data gathered on the magnitude of this effect vary considerably. Gates *et al.* [3] mention data indicating reaction rate reductions of 10 to 15% for each mole (or volume) percent H_2S in the reactant gas mixture. Other reports have indicated activity reductions per percentage H_2S ranging from less than 3 to over 50%.

The reason for this wide range is, not surprisingly, a high dependence

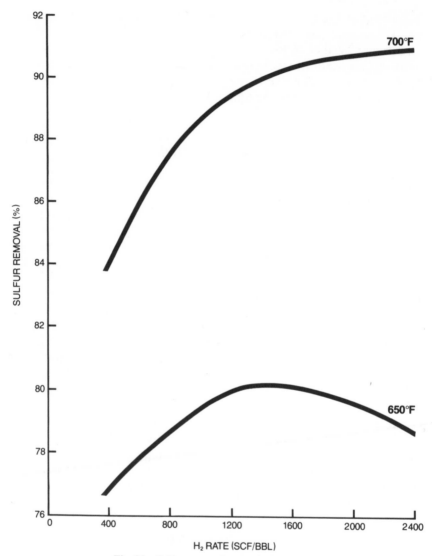

Fig. 22. Sulfur removal versus treat gas rate.

upon specific feedstock properties and operating conditions. In general, less of an effect is seen on heavier, high-sulfur feeds and more of an effect on lighter, low-sulfur feeds. A typical effect on a moderate sulfur level, moderate severity, middle-distillate hydrotreater might be a 5–10% activity loss for each volume percent H_2S in the total treat gas to the reactor. For a typical

H₂S level of, say, 5% in the treat gas, this calculates to a loss of roughly 25 to 50% (although this activity loss is not really linear).

Many, if not most, refiners find that this inhibition effect can justify the capital and operating costs required for a recycle gas scrubber to remove H₂S.

E.   MERCAPTAN RECOMBINATION PROBLEM

In naphtha hydrotreaters, many refiners have reported the problem of olefins and H₂S reacting to form mercaptans after exiting the catalyst bed, so that the catalyst has no chance to desulfurize the mercaptan. Even though 1 or 2 ppm sulfur may be involved, this is still too high for good reformer feedstock.

Because the thermodynamic equilibrium among H₂S, olefins, and mercaptans favors some conversion to mercaptans, the amount of olefins exiting the hydrotreating reactor must be minimized. This may be particularly difficult if some of the olefins are being formed in the reactor by hydrocracking, rather than just unconverted olefins in the feedstock. There is no clear-cut solution to this problem, but the following observations may be made:

(1)   Several refiners have solved this problem by lowering reactor temperatures, thus slowing down the reaction of olefins plus H₂S to mercaptans (and possibly decreasing olefin formation by hydrocracking). Temperatures above 650°F have been found to be especially troublesome.

(2)   The problem is more common with increasing amounts of cracked streams (and thus olefins) in the feed.

(3)   The problem is the worst in units with long residence times at high temperatures after exiting the catalyst bed.

(4)   The use of a recycle gas scrubber to minimize H₂S levels can be a big help.

F.   PRESSURE DROP

The most common cause of premature termination of a commercial operating run is a high pressure drop across the hydrotreating reactor. In most cases, the problem is not an initial clean bed pressure drop (which can be estimated reasonably well using various equations applicable to all vapor or vapor–liquid flow through packed beds), but rather a pressure drop buildup as the run progresses. Several common causes and corrective measures are as follows.

1. Inorganic Particulate Matter Entering the Reactor

This is perhaps the most frequent cause of pressure drop buildup problems. Iron sulfide scale is the biggest contributor. Typical preventive measures include trash baskets, guard reactors, and inlet feed filters, as discussed in Section V.C. Some refiners have successfully used liquid cyclones to remove particles from the feed, although these tend to work well only on relatively coarse particles. Either of the following after-the-fact corrective measures is primarily used:

(a)   Skimming the top of the reactor—this can be done either before (in which case inert entry conditions prevail) or after regeneration, generally by vacuuming off the upper 2–6 ft of support balls and catalyst bed and then replacing with fresh material. This procedure works well when particulate contamination is concentrated at the top of the reactor, which usually is the case.

(b)   Dumping, screening, and reloading the entire reactor—also was discussed in Section V.C. Some refiners have successfully used magnets to remove iron particulates during this dumping and screening operation.

2. Organic Particulate Matter

This normally is a polymerization product resulting from the combination of olefins (especially diolefins), oxygen, temperature, and time. These polymers, or gums, can be formed in the reactor but more commonly are formed upstream. Typical preventative measures are

(a)   Minimizing residence time from the upstream unit to the hydrotreating reactor, especially for thermally cracked stocks such as coker naphtha. Whenever possible these streams should be run hot to the hydrotreater.

(b)   When intermediate storage is necessary, minimizing contact with oxygen by the use of floating roof tanks, inert gas blanketing, etc. Also, polymer formation may occur if an oxygen-free olefinic feed is mixed with a thermally stable SR feed that has been exposed to air. Any significant polymer buildup invariably results in premature shutdown and regeneration.

3. Inorganic Fines Generated within the Reactor

Although less commonly the cause of pressure drop problems than inorganic particles entering with the feed, fines can be generated within the reactor by catalyst breakup. Typical preventative measures are

(a)   Gentler catalyst handling during loading—Each catalyst supplier has specific recommended procedures, although most commercial catalysts today can withstand reasonably rough handling.

(b)   Minimizing catalyst bed movement—Most catalyst fines probably are created by attrition rather than actual crushing. Support balls on the top of downflow reactor beds, inlet flow distributors, etc., help to keep catalyst beds in place. Probably the worse creation of catalyst fines occurs from flow reversals.

(c)   If a catalyst is ever wet with water, a slow heat-up is required to avoid catalyst fracturing by rapid steam generation.

Once significant catalyst fines have been generated, dumping and screening of the bed are required.

G.   APPARENT ACTIVITY DECLINE

The causes and results of catalyst deactivation have been discussed in Section V.E. Although it is evident that many different factors cause catalysts to deactivate, apparent activity decline on a commercial hydrotreating unit frequently is not related to catalyst deactivation at all. Let us consider two common examples.

1.   Channeling

In any fixed bed reactor, any maldistribution of the inlet feed, any uneven packing of the catalyst bed, and/or any blockage of the void space between catalyst particles results in a degree of uneven flow distribution and, therefore, less than optimum feed–catalyst contact. Especially in the case of mixed phase trickle bed reactors, liquid feed naturally seeks out relatively open "channels" that offer the least resistance to flow.

Channeling can have a significant effect on unit performance well before it is evidenced by a high pressure drop. In such cases, the tendency is to assume that unit performance is decreasing as a result of catalyst deactivation, instead of realizing that some of the catalyst bed, albeit very active, is not effectively contacting the feed.

One standard method for identifying channeling is to increase the unit temperature and check whether the product sulfur level decreases as expected—if not, this is a good indication of channeling. Corrective measures are as discussed above for pressure drop.

2.   Heat Exchanger Leaks

Feed–effluent heat exchanger leaks probably have been responsible for many cases of rapid apparent activity decline over the years, but such leaks, if small, have been very hard to detect. However, the need to reduce sulfur and nitrogen levels to 0.5 ppm maximum, or even lower, on feed to bimetallic reforming catalysts has made it imperative to eliminate all such leaks in naphtha hydrotreaters. Obviously, a leak of only 1 part feed into 500 parts effluent makes it impossible to reduce a 500-ppm sulfur naphtha to a below 1-ppm sulfur product.

A good indication of feed and effluent exchanger leakage is the temperature response test described for the detection of channeling, since any sulfur compound bypassing the hydrotreating reactor completely will obviously not be reacted. Other more definite detection methods include the injection of gasoline dyes or radioactive tracers. Grossman [23] describes the use of a helium leak test.

H. COLOR

One of the more perplexing observations in hydrotreating is the occasional appearance of color bodies in the product, frequently reported as red or orange and usually in middle-distillate fractions, especially those derived from cracked stocks.

There is no single commonly accepted explanation for this problem, but most people believe the color bodies to be higher-boiling, condensed ring compounds containing nitrogen. Formation of these color bodies appears to be favored by low space velocities (especially related to poor flow distribution, thus creating localized long residence times), low H_2 partial pressures, and high temperatures.

VII. Economics

A. CAPITAL AND OPERATING COSTS

Typical ranges of hydrotreating costs (in 1981 U.S. dollars) are estimated in Table VII. Capital investment includes all battery limits capital through product fractionation. Total operating costs include all capital-related charges, labor, utilities, chemicals and catalysts, and a hydrogen consumption charge of $2/kSCF.

Catalyst costs normally are not a significant portion of either the capital or operating expense, except for residual oil hydrotreating. As an example, a typical catalyst life of 10 bbl/lb in atmospheric resid service at a price of, say, $5/lb calculates to a 50¢/bbl operating cost.

TABLE VII

Typical Ranges of Hydrotreating Costs in 1981 U.S. Dollars

Application	Capital investment ($/BPSD capacity)	Total operating costs ($/barrel feed)
Naphtha	200–400	0.15–0.30
Middle distillate	300–600	0.30–0.60
VGO	600–1000	0.60–1.50
Atmospheric resid	1500–3000	2.50–5.00

B. VALUE OF PRODUCT IMPROVEMENTS

The value of product improvement as a result of hydrotreating can be divided into two basic categories.

1. Meeting Product Sales Specifications

Specifications such as product sulfur level, cetane index, and smoke point frequently can be met only by some degree of hydrotreating. In such cases, the value of hydrotreating is equal to the incremental value of selling the product versus the values and costs of alternate uses.

2. Downstream Processing Benefits

Naphtha hydrotreating to provide good reformer feed is easily justified and essentially universally practiced, with the exception of only a few units that still have very low-sulfur (only a few ppm) feeds and use monometallic reforming catalysts. Many refiners find catalytic feed hydrotreating economically attractive. Ritter *et al.* [9] presented an example where an $8 – 10 million investment yielded incremental earnings of $4.50 to 6.00 million/yr.

C. FUEL SAVINGS WITH CATALYST ACTIVITY

With the rapid increase in fuel costs over the last decade, higher-activity hydrotreating catalysts have become increasingly popular because of their ability to do a given job at a significantly lower temperature. NPRA [24] discussed this subject at some length, with activity gains of between 20 and 50°F reported for switching to higher-activity catalysts. Taking into account furnace and heat exchanger efficiencies, the energy savings resulting from these lower temperature operations were estimated at 90 Btu/bbl of feed per degree Fahrenheit.

In addition to direct fuel savings, lower temperature operations offer the harder-to-quantify but still very real benefits of longer run lengths and decreased corrosion rates.

VIII.   Future Outlook

Hydrotreating has become, and unquestionably will remain in the foreseeable future, a major processing scheme for any modern refinery. However, it is hard to predict exactly how fast the use of hydrotreating will grow.

Perhaps the major reason for this uncertainty arises from the many alternate processes being considered for converting residual fractions and heavy crudes to clean and valuable liquid products. Many articles have been written about this subject recently, such as Whittington et al. [11] and Sosnowski et al. [25]. There is no doubt that the crude slate available worldwide is becoming increasingly heavier, more sour, and of higher metals content, while at the same time the demand for heavy fuel oils (even low-sulfur fuel oils) is steadily decreasing.

This combination of events creates a great need for increased conversion and upgrading of the "bottom of the barrel." Essentially all the alternate process schemes being considered involve hydrotreating, but the amount and severity of hydrotreating required is highly dependent upon answers to questions such as, How much resid converson is done thermally versus catalytically? As usual, individual refinery economics will resolve these questions.

Another major factor affecting the growth of hydrotreating is the future of the synthetic fuels industry, specifically direct catalytic conversion of coal and downstream upgrading of liquids from coal, shale, and tar sands. Since these energy sources are highly hydrogen-deficient compared to conventional crude oils, large amounts of hydrogen must be added to these molecules somehow. In addition, the levels of nitrogen, oxygen, and some metals (especially arsenic in shale oil) are much higher than those in crude oil, so relatively more hydrotreating is required.

At the present time, Canadian tar sands operations already have been producing upgraded liquid products for several years, and the first commercial shale oil operation is due on stream in 1983. Several coal liquefaction (direct and indirect) and gasification pilot operations are underway, but the commercial economics and timing of these processes are not yet well defined.

References

1. L. R. Aalund and A. Cantrell, *Oil Gas J.* **79**, 13, 63 (1981).
2. L. Auldridge and A. Cantrell, *Oil Gas J.* **78**, 52, 75 (1980).
3. B. C. Gates, J. R. Katzer, and G. C. A. Schuit, "Chemistry of Catalytic Processes." McGraw–Hill, New York (1979).
4. S. K. Alley, Union Oil Company of California, First International Unicracking—Unicracking/HDS Conference, January (1981).
5. C. J. Keating, and J. B. MacArthur, *Hydrocarbon Process.* **59**, 12, 101 (1980).
6. R. H. Hass, M. N. Ingalls, T. A. Trinker, B. G. Goar, and R. S. Purgason, *Hydrocarbon Process.* **60**, 5, 104 (1981).
7. NPRA Brochure on Lubricating Oil and Wax Capacities. National Petroleum Refiners Association (1980).
8. D. C. McCulloch, *Oil Gas J.* **73**, 29, 53 (1975).
9. R. E. Ritter, J. J. Blazek, and D. N. Wallace, *Oil Gas J.* **72**, 41, 99 (1974).
10. A. G. Bridge, G. D. Gould, and J. F. Berkman, *Oil Gas J.* **79**, 3, 85 (1981).
11. E. L. Whittington, V. E. Pierce, and B. B. Bansal, *Chem. Eng. Prog.* pp. 45–50, February (1981).
12. J. R. Murphy, and S. A. Treese, *Oil Gas J.* **77**, 26, 135 (1979).
13. E. S. Ellis, J. Sosnowski, R. L. Hood, M. G. Luzarraga, and K. L. Riley, Extension of Residfining Technology to Hydroconversion. Paper at API Midyear Refining Meeting, Chicago, May 13 (1981).
14. R. L. Richardson, F. C. Riddick, and M. Ishikawa, *Oil Gas J.* **77**, 22, 80 (1979).
15. NPRA Question and Answer Session on Refining and Petrochemical Technology. National Petroleum Refiners Association, pp. 128–129, October (1980).
16. A. I. Snow and M. P. Grosboll, *Oil Gas J.* **75**, 21, 61 (1977).
17. J. J. Carberry, "Chemical and Catalytic Reaction Engineering." McGraw–Hill, New York (1976).
18. M. A. Christ, G. N. Shah, and L. G. Sherman, *Oil Gas J.* **77**, 22, 95 (1979).
19. W. C. van Zijll Langhout, C. Ouwerkerk, and K. M. A. Pronk, *Oil Gas J.* **78**, 48, 120 (1980).
20. K. Shuke, *Petrotech* **4**, 2, 147 (1981).
21. D. C. McCulloch, and R. A. Roeder, *Hydrocarbon Process.* **55**, 2, 81 (1976).
22. M. D. Edgar, *Oil Gas J.* **76**, 33, 102 (1978).
23. A. P. Grossman, *Hydrocarbon Process.* **54**, 1, 58 (1975).
24. NPRA Question and Answer Session on Refining and Petrochemical Technology. National Petroleum Refiners Association, p. 120, October (1978).
25. J. Sosnowski, D. W. Turner, and J. Eng, *Chem. Eng. Prog.* pp. 51–55, February (1981).

CHAPTER 5

Catalytic Reforming of Naphtha in Petroleum Refineries

M. DEAN EDGAR

Catalyst Department
American Cyanamid Company
Houston, Texas

I. Introduction

A. DEFINITION OF CATALYTIC REFORMING

Catalytic reforming is a refinery process in which a naphtha feed (C_5–400°F) is passed through several reactor beds of catalyst at high temperature and moderate pressure to achieve an increase in the aromatic content of the naphtha or an increase in its octane number. Normally the naphtha has been hydrotreated to remove impurities that either inhibit the reactions or poison the reforming catalyst. The naphtha can be obtained directly from the crude unit or from the fractionated product of another refinery process, such as a coking. The catalyst is generally a few tenths percent platinum (in admixture with other noble metals and a halogen) supported on a pure alumina base.

B. HISTORY

The catalytic reforming process was developed during 1947–1949 [1]. In the next 7 yr, 13 new commercial reforming processes were developed and licensed by various petroleum and engineering companies [2]. The process has continued to evolve through the years, the latest development being a design in which the catalyst moves continuously through the reactor, from reactor to reactor, and finally to a regeneration vessel.

In addition to changes in the process design, the catalyst used in reforming has been modified to offer improved performance. The most significant development, which occurred in the late 1960s, is the bimetallic (platinum–rhenium) reforming catalyst.

According to data reported in the *Oil and Gas Journal* [3] as of January 1, 1981, U.S. reformer capacity was 4,051,400 barrels/stream day (bbl/sd). Seventy-five percent of these units used a bimetallic reforming catalyst, whereas the remaining 25% used a straight platinum catalyst. This represents about 290 reforming units in the United States.

C. PURPOSE OF THIS CHAPTER

The intent of this chapter is to present an overview of the reforming process. It will cover feedstock properties, reactions, process descriptions, catalysts, and operating variables.

II. Feed Components and Reactions

The hydrocarbon feed to the reformer is usually a depentanized stream with a 400°F maximum ASTM D-86 distillation end point. This feed can originate from a crude distillation unit or from the fractionation of products from another process unit such as a coker. Hydrocarbons in the feed that have fewer than six carbon atoms (C_5-) are not considered to be involved in the reactions. It is desirable to remove them from the feed, since their presence physically interferes with the access of the reformable hydrocarbons to active sites on the catalyst.

A. HYDROCARBON TYPES

Besides carbon number, the components of the feed are grouped by the following types: paraffins, naphthenes, and aromatics.

Paraffins are saturated straight- or branched-chain hydrocarbon molecules. Straight-chain molecules are called normal paraffins, and branched-chain molecules are referred to as isoparaffins. Naphthenes are saturated ring compounds that may have side chains attached to the ring. Aromatics are ring compounds in which the carbon atoms are bonded by resonating single and double bonds. A six-carbon benzenoid ring is the basic aromatic structure to which side chains or other rings may be attached. Unsaturated hydrocarbon compounds, classed as olefins, react rapidly with the reformer catalyst to form coke in the reformer. Generally these compounds are saturated during the feed preparation step in the hydrotreater which normally precedes the reformer.

B. REACTIONS

The main reactions that occur during the reforming process are naphthene dehydrogenation, naphthene isomerization, dehydrocyclization, paraffin isomerization, and hydrocracking. Examples of each of these reactions are shown in Fig. 1.

1. Naphthene Dehydrogenation

Naphthene dehydrogenation is a relatively fast reaction in which naphthenes are converted to aromatics. Most of the naphthene dehydrogenation is completed in the first reactor of the reformer. Because this reaction is highly endothermic, there is a substantial reduction in temperature across

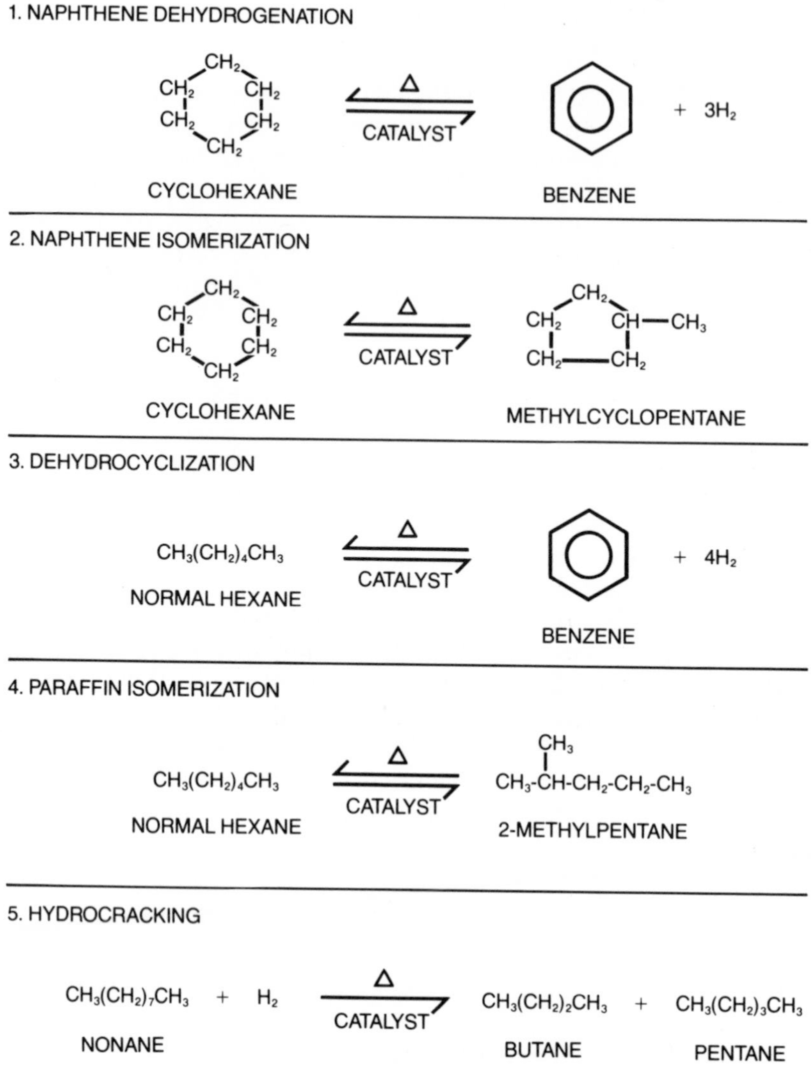

1. NAPHTHENE DEHYDROGENATION

CYCLOHEXANE   BENZENE   + 3H$_2$

2. NAPHTHENE ISOMERIZATION

CYCLOHEXANE   METHYLCYCLOPENTANE

3. DEHYDROCYCLIZATION

CH$_3$(CH$_2$)$_4$CH$_3$

NORMAL HEXANE

+ 4H$_2$

BENZENE

4. PARAFFIN ISOMERIZATION

CH$_3$(CH$_2$)$_4$CH$_3$

NORMAL HEXANE

$$CH_3\text{-}\overset{\overset{\displaystyle CH_3}{|}}{C}H\text{-}CH_2\text{-}CH_2\text{-}CH_3$$

2-METHYLPENTANE

5. HYDROCRACKING

CH$_3$(CH$_2$)$_7$CH$_3$ + H$_2$ CH$_3$(CH$_2$)$_2$CH$_3$ + CH$_3$(CH$_2$)$_3$CH$_3$

NONANE   BUTANE   PENTANE

Fig. 1. Examples of reforming reactions.

the first reactor. Temperature decreases in excess of 100°F are common for a midcontinent-type naphtha. This reaction is catalyzed by the precious metal portion of the catalyst. This reaction produces hydrogen, and its rate is slowed by high hydrogen partial pressures. The conversion of naphthenes to aromatics produces an increase in product density.

2. Naphthene Isomerization

Naphthene isomerization reactions proceed quickly by action with both the acidic (halogen) portion of the catalyst and, to a lesser degree, the precious metal portion of the catalyst. This reaction produces a rearrangement of the molecule with no addition or loss of hydrogen; therefore, the reaction rate is virtually unaffected by pressure. The exothermic temperature effects associated with naphthene isomerization are usually small enough to go undetected in a commercial reforming unit.

3. Dehydrocyclization

Dehydrocyclization, an important octane-enhancing reaction in which paraffins are converted to aromatics, is a relatively slow reaction catalyzed by both the precious metal and the acid portions of the catalyst. This endothermic reaction usually occurs in the middle to the last reactors of the reformer unit. Dehydrocyclization produces hydrogen, and its rate is inhibited by high hydrogen partial pressure. Dehydrocyclization reactions increase the density of the product.

4. Paraffin Isomerization

Paraffin isomerization is a relatively fast reaction catalyzed mainly by the acid function of the catalyst. Like naphthene isomerization, this reaction produces a rearrangement of the molecular structure with no net change in hydrogen production. The rate of paraffin isomerization is not strongly affected by hydrogen partial pressure. Exothermic temperature effects associated with paraffin isomerization are not usually detected in a refinery reformer.

5. Hydrocracking

Hydrocracking, breaking long-chain paraffins into smaller-chain paraffins, is mostly catalyzed by the acid function of the catalyst. This relatively slow reaction is generally undesired, since it produces excessive quantities of light ends — C_4 and lighter hydrocarbons — and coke and consumes hydrogen that could be used elsewhere in the refinery. The rate of hydrocracking is enhanced by high unit pressure. Hydrocracking is exothermic and normally occurs in the last reactor. In some cases enough hydrocracking occurs to produce a temperature increase across the last reactor. Hydrocracking reactions reduce the density of the product.

The preceding was a general description of reformer reactions. A more detailed treatment of reformer reactions is contained in the book by Gates et al. [4].

C. Feedstock

In Table I are three examples of naphtha feeds. The composition of the feed has a major effect on the reactor temperature required to achieve a desired product octane value and on the quantity of reformate yield obtained. The higher the paraffin content of the feed, the harder it is to reform. A high-paraffin naphtha, such as light Arabian naphtha, requires higher reactor temperatures, produces less reformate, and causes shorter cycle lengths than the other naphthas shown in Table I. There are various ways to rank naphthas as to reformability. One method uses a value determined by adding the naphthene content to twice the aromatic content and is referred to as the N + 2A value. As shown in Table I, N + 2A values of 60–65 are typical of naphthas from midcontinent crudes, which are similar to the naphthas being run by U.S. refiners until the late 1970s. As refiners moved into the 1980s and were required to use less desirable crudes, the N + 2A values of naphthas began to decrease. Naphthas similar to the light Arabian naphtha in N + 2A value are becoming more common.

Table II illustrates the differences in properties between feed and product for a midcontinent-type naphtha reformed to 95 clear research octane number (RONC). As can be seen, there is a substantial increase in the aromatic content of the reformate at the expense of the napthenes. Since the paraffins are difficult to reform, there is only a small reduction in the paraffin content between feed and reformate.

TABLE I
Feedstock Examples

Naphtha properties	West Coast naphtha	Midcontinent naphtha	Light Arabian naphtha
Gravity (°API)	49	56	66
ASTM D-86 distillation (°F)			
IBP	230	178	176
10%	272	221	194
30%	287	235	205
50%	302	246	216
70%	326	264	232
90%	350	289	254
EP	385	330	298
Composition (vol%)			
Paraffins	22	45	74
Naphthenes	56	45	19
Aromatics	22	10	7
	100	100	100
N + 2A	100	65	33

TABLE II
Feed versus Reformate Comparison for Midcontinent Naphtha

Properties	Feed	Reformate
Gravity (°API)	56	46
ASTM D-86 distillation (°F)		
IBP	178	140
10%	221	200
30%	235	230
50%	246	245
70%	264	270
90%	289	305
EP	330	365
Volumetric average boiling		
point (°F)	251	250
Reid vapor pressure (psia)	1.0	2.5
Composition (vol%)		
Parafins	45	40
Naphthenes	45	5
Aromatics	10	55
	100	100
Octane (RONC)	55	95

III.   Process Description

A.   UNIT CLASSIFICATION

Reforming units are usually classified as belonging to one of the following three categories: semiregenerative, cyclic, or moving bed. These classifications reflect the manner and frequency of regeneration of the reforming catalyst.

1.   Semiregenerative Units

A block diagram of a semiregenerative unit is shown in Fig. 2. Pretreated feed enters the unit through a feed–effluent heat exchanger. From the exchanger, the feed, combined with the recycle gas, goes to a heater to increase its temperature to a range of about 900–980°F. The feed then goes through three or four reactors in series. Since the reforming reactions affecting the reactor temperature are mainly endothermic, there are furnaces between the reactors to restore the temperature of the feed stream to the desired level. The product from the last reactor goes through the feed–effluent heat exchanger and then to a flash drum. At the flash drum,

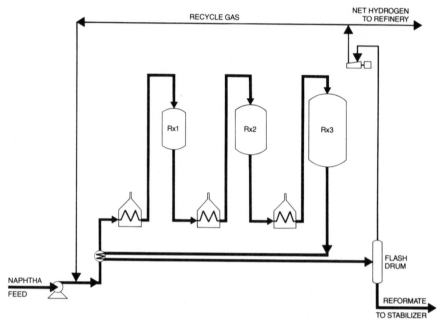

Fig. 2. Example of a semiregenerative reforming unit. Rx, Reactor.

the liquid product is taken off the bottom of the vessel, and the overhead is divided into product hydrogen and recycle gas. The liquid product is sent to a stabilizer to remove the light ends.

These units have to shut down periodically to regenerate the catalyst which becomes deactivated as a result of coke deposition. Generally refiners strive to operate these units to achieve at least a 6-month time interval between regenerations. This limits the maximum product octane value obtained to about 100 RONC.

2. Cyclic Units

Figure 3 is an example of a cyclic unit. It differs from a semiregenerative unit in that it has one additional reactor and a manifold system that allows the catalyst in one reactor to be regenerated while the catalyst in the other reactors is processing feed. The additional reactor is known as a swing reactor. The swing reactor can be substituted for any of the series reactors. This design can take advantage of low unit pressures to gain a higher C_5+ reformate yield and hydrogen make and can be operated at high octane levels (100+ RONC) that would result in unacceptably short cycle lengths in

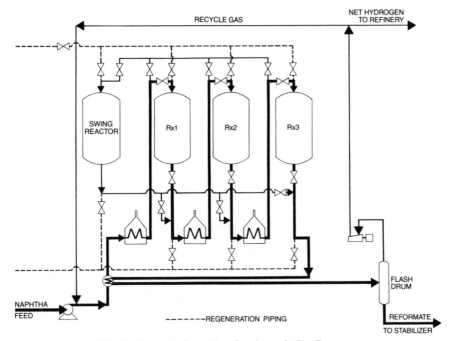

Fig. 3. Example of a cyclic reforming unit. Rx, Reactor.

a semiregenerative unit. Coke make is usually higher in the last reactors of a reformer because of their higher average bed temperature. Therefore reactors in these positions in a cyclic unit are normally regenerated more frequently than the first reactors. In addition to allowing operation at higher severity or lower pressure than in a semiregenerative unit, the more frequent regeneration of individual reactors in a cyclic unit results in less of a decline in C_5+ reformate yield and hydrogen production with time on stream when compared to the yields of a semiregenerative unit.

3. Moving Bed Units

A moving bed unit, shown in Fig. 4, is an extension of the cyclic unit concept. In the most common moving bed design, the reactors are stacked one atop the other except for the fourth (last) reactor which frequently is set beside the other stacked reactors. The flow path of the feed is similar to that of the other reformer designs in that the feed is heated by exchange, heater and interheaters, and exits through a flash drum en route to a stabilizer. The reactors are radial flow in design. The catalyst is slowly moved from the first

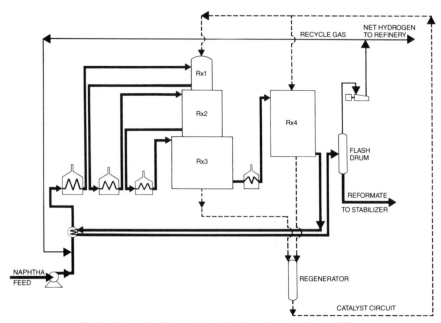

Fig. 4. Example of a moving bed reforming unit. Rx, Reactor.

(top) reactor to the bottom reactor. The coked catalyst is sent to the regeneration section. Catalyst flows through the fourth reactor, if present, as a separate system. These units can be built without the regeneration section and operated as semiregenerative units.

A less common design for a moving bed reformer has the individual reactors placed separately, as in typical semiregenerative fashion, with provisions for moving the catalyst from the bottom of one reactor to the top of the next reactor in line. Coked catalyst is withdrawn from the last reactor and sent to a regeneration vessel. Fresh catalyst or regenerated catalyst is added to the top of the first reactor to maintain a constant quantity of catalyst.

Because there is a mechanism that prevents excessive coke buildup on the catalyst, these units can operate at low pressures, e.g., 100 psig, and/or high severity, e.g., 100+ RONC. Reformate yield loss or hydrogen production decline over time on stream is minimized by selecting the correct catalyst circulation rate.

B.   PURPOSE OF REFORMING

Reformers are operated primarily to produce either motor fuel or aromatics. In either case, hydrogen is produced that can be used in other refinery units such as hydrotreaters.

Motor fuel production usually uses a full range or heavy naphtha with an end boiling point of about 400°F. The octane of this naphtha is increased to a 95 – 102 RONC range. This material provides the high-octane component that is blended with other refinery streams boiling in the gasoline range to produce the finished product.

For aromatics production, frequently a light naphtha feed with an end boiling point of 310 to 340°F is sent to the reformer. These units are called BTX units (for benzene, toluene, xylene). The product from the BTX reformer is normally sent to an aromatics extraction unit. The benzene, toluene, and xylenes are utilized as raw materials for various petrochemical processes.

C.   UNIT DESIGN VARIABLES

Regardless of its classification when a reforming unit is in the design stage, there are several variables to consider: liquid hourly space velocity (LHSV); pressure; hydrogen-to-hydrocarbon molar ratio ($H_2:HC$); and type, number, and size of reactors.

1.   Liquid Hourly Space Velocity

Common values of LHSV range from 1 to 3 hr^{-1}. For a unit of a specified capacity, selection of the LHSV determines the volume of catalyst required. Except for very low values of space velocity, e.g., less than 1.0, which tend to favor hydrocracking, the magnitude of the space velocity's effect on yield selectivity is considerably less than the effect of other design variables. The primary effect of space velocity on unit operation is the cycle length obtained and the start-of-run temperature required to achieve a desired product octane level.

2.   Pressure

The pressure selected for the operation has a major effect on yield and cycle length. Pressures of 400 to 500 psig favor a long cycle length. However, by reducing the pressure, the dehydrogenation reaction equilibrium is shifted in a direction favoring increased aromatics yield and hydrogen production. The decrease in unit pressure reduces the likelihood of hydro-cracking. Enhancing the dehydrogenation reactions and inhibiting the hy-drocracking reactions result in an increase in the C_5+ reformate yield. Unfortunately a reduction in pressure also increases the rate of coke deposi-tion on the catalyst, which reduces the cycle length. Semiregenerative units could not take advantage of low-pressure operation until the advent of bimetallic and multimetallic reforming catalysts with their ability to tolerate higher coke levels. Pressures of 200 to 250 psig could then be used while obtaining cycle lengths similar to those obtained with straight platinum catalysts at higher pressures. With cyclic units and moving bed units, pressures as low as 85 psig have been considered [4].

3. H_2:HC Molar Ratio

The H_2:HC molar ratio is a measure of the recycle hydrogen flow rate. The trend has been to reduce the H_2:HC molar ratio from $8:1-10:1$ to $3:1-5:1$. A reduction in the H_2:HC ratio reduces the compressor needs of the unit. However, a reduction in the H_2:HC ratio increases the rate of coke make, which reduces the cycle length. For example, a reduction in the H_2:HC ratio from $5:1$ to $4:1$ reduces the cycle length by about 20%, all other conditions remaining the same. A reduction in the H_2:HC ratio reduces the hydrogen partial pressure and, like a reduction in unit pressure, enhances the dehydrogenation reaction and inhibits the hydrocracking reaction. The magnitude of the effect on yields of a reduction in the H_2:HC ratio from $8:1-10:1$ to $3:1-5:1$ is not as great as the effect of a reduction in pressure from 500 to 200 psig.

4. Reactors—Type, Number and Size

The reactors used in reforming units are classified as either downflow or radial flow. These types of reactors are illustrated in Fig. 5. In a downflow reactor, feed enters at the top and flows downward through the catalyst bed. The product exits at the bottom. In a radial flow reactor, feed enters at the top and product exits at the bottom, but the feed flows across an annular catalyst bed to a center pipe. Reforming units are comprised of all radial flow reactors, all downflow reactors, or a combination of both.

The advantages of the radial flow reactor design are a low pressure drop across the reactor and a low fouling rate of the catalyst bed due to particu-lates carried in with the feed. In units in which there is concern about scale carryover into the first reactor, the first reactor is frequently of the radial flow design. The disadvantages of the radial flow reactor design are problems in characterizing the flow pattern of the feed to obtain full utilization of the catalyst and possible bed settling which allows feed to bypass the catalyst.

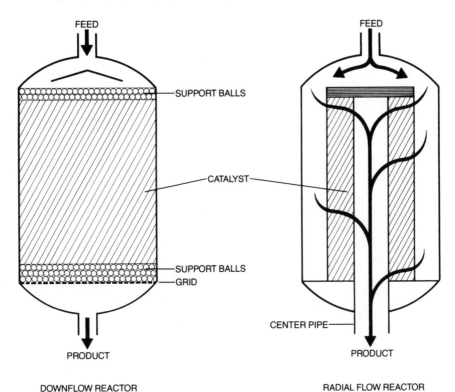

Fig. 5. Examples of reactor design.

Semiregenerative reformers are generally built with three to four reactors in series. Initially the units had three reactors. A fourth reactor was added to some units to allow an increase in either severity or throughput while maintaining the same cycle length. At constant octane operation there does not appear to be a significant yield advantage for four reactors over three reactors when the addition of the fourth reactor reduces the space velocity [6].

The size of the first reactors is small compared to the size of the end reactors. The very rapid endothermic reactions that occur in the first reactors can reduce the reactor temperature to a point at which further reaction stops. Additional catalyst in the bed below this point is not utilized effectively. Thus the small reactor size is favored. The reactions that occur in the last reactors are relatively slow, so large reactors are required. A typical reactor size distribution is shown in the accompanying tabulation.

Three-reactor system		Four-reactor system	
Reactor no.	Percent	Reactor no.	Percent
1	20	1	12
2	30	2	20
3	50	3	28
		4	40

Sometimes, when a semiregenerative reformer's capacity is expanded, two existing reactors are placed in parallel, and a new, usually smaller, reactor is added. Frequently the parallel reactors are placed in the terminal position. When evaluating unit performance, the parallel reactors are treated as though they are a single reactor of equivalent volume.

Cyclic units typically use five or six reactors including the swing reactor. Moving bed units generally use three or four reactors.

The preceding items are set at the time of unit design as a series of compromises that enable the unit to handle the range of feedstock types envisaged to produce the desired product quality and to provide an acceptable cycle length.

IV. Catalysts

A. DUAL FUNCTION

Modern reforming catalysts are dual function catalysts. One function is the hydrogenation–dehydrogenation function, and the other is the acid function. As indicated in Section II.B, some reactions require one or both of these functions. For the catalyst to provide the desired products under reforming conditions, the two catalyst functions must be balanced. The balance must be maintained throughout the cycle to obtain the best performance. The severity of the operation affects the desired balance between the two functions.

B. SUBSTRATE FORM

Reforming catalysts have evolved over the years along with advances in the process itself. The most commonly used reforming catalysts today consist of one or more precious metals on an alumina support. The alumina used for support is one of two crystalline forms—eta or gamma. The eta form has a higher acid function than the gamma form. The eta form has served as the support for mainly straight platinum catalysts. η-Alumina is characterized by a high initial surface area. Following use and regeneration, the surface area begins to decline. Total life is limited by this loss in surface area to just a few cycles.

γ-Alumina does not have as high an acid function as η-alumina, but it is more thermally stable and retains more of its initial surface through repeated use and regeneration than the eta form. γ-Alumina-based catalysts used in cyclic reformers may undergo several hundred regenerations before losing enough surface area to require replacement. The lower acid function of the γ-alumina catalyst can be compensated for by proper adjustment of the halogen content of the catalyst.

C. PHYSICAL PROPERTIES

The physical properties of reforming catalysts, regardless of manufacturer, tend toward similar ranges. The surface area measured by instrumental analysis is about $175-300$ m²/g. The pore volume measured by water ranges from 0.45 to 0.65 cm³/g. The catalyst particles are either extrudates or spheres ranging from $\frac{1}{16}$ to $\frac{1}{12}$ in. diameter. The typical crush strength of these catalysts is about $3-7$ lb/mm. The density of reforming catalysts ranges from ~ 32 to ~ 49 lb/ft³.

D. PROMOTER METALS

Various promoter metals have been used in reforming catalysts. Platinum gained widespread use in the 1950s and early 1960s. In the middle to late 1960s, the addition of rhenium to platinum-containing catalysts began to see commercial use. The function of rhenium and its form on a catalyst are still debated, but it appears to increase the catalyst's coke tolerance. This allows refiners to reduce the pressure or H_2:HC ratio or to increase operating severity while maintaining the same cycle length obtained with the straight platinum-promoted catalyst.

The patent literature contains references to the incorporation of other metals, such as tin, germanium, and lead, onto the platinum-supported catalyst. Catalysts containing platinum and incorporating other metals have been used in commercial units. Promoter metals usually comprise 1 wt% or less of the finished reforming catalyst.

E. BENEFITS OF BIMETALLIC CATALYSTS

The benefits of bimetallic catalysts are primarily related to their enhanced coke tolerance. Bimetallic reforming catalysts allow operations to continue until levels of 20 wt% coke on the catalyst in the last reactor are reached. The better coke tolerance has been used mainly to allow a reduction in reactor pressure and thereby to obtain a yield advantage for new and revamped units. Other ways to take advantage of the enhanced coke tolerance are a reduced H_2:HC ratio, increased feed rate, increased product octane or aromatics yield, and reduced catalyst volume in new unit designs.

F. START-UP AND PRESULFIDING

Fresh reforming catalyst readily picks up moisture. This moisture can be from rain or high humidity present during loading or from water that has collected in the low spots in the unit. Because water removes some of the chloride placed on the catalyst, this could upset the balance between the precious metal function and the acid function. Therefore care should be taken to dry the unit, drain low spots, and load the catalyst under conditions as dry as possible.

Reforming catalysts are supplied with precious metals either in the oxide form or already in the reduced and presulfided form. If the catalyst is in the oxide form, it must be reduced and, especially for bimetallic catalysts, presulfided before feed is introduced into the reformer unit.

After loading the catalyst, the unit is closed and pressure-tested with a nitrogen atmosphere. After determining that the unit is leak-free, heating of the catalyst begins. A good practice is to maintain a temperature differential among the reactors of 50°F between the first and last reactor, with the last reactor having the highest temperature. This prevents condensation of any water driven off during heating from occurring in a downstream reactor. When the reactor temperatures line out at about 700–800°F, reduction of catalyst in the oxide form takes place when the nitrogen atmosphere is displaced by high-purity hydrogen.

The catalyst, especially a bimetallic catalyst, is in a highly reactive state following reduction. If feed is introduced at this point, methanation reactions and excessive hydrocracking are likely to occur. These reactions are exothermic and could result in a temperature runaway, with subsequent damage to the catalyst. To prevent this from occurring, the refiner should temporarily deactivate the catalyst; most commonly this is done with sulfur. Following reduction, a typical procedure might call for 0.06 wt% (based on the weight of the catalyst) sulfur to be introduced into the individual reactors. Hydrogen sulfide has been widely used for this purpose, although some refiners prefer dimethyl sulfide because of easier handling. After injection of the sulfiding agent is completed, the recycle gas stream is checked for the presence of H_2S as a positive indication that the sulfur has worked its way through the catalyst beds.

Once the catalyst has been reduced and presulfided, it is ready to accept the feed. Sweet naphtha is introduced at reactor bed temperatures ranging from 700 to 850°F. Use of the lower temperature when feed is introduced may require more time to achieve the desired octane, but it reduces the chances of a temperature runaway. There is also some latitude in the reactor temperatures when feed is introduced, depending on the naphthenic content of the feed. The large temperature drop due to a high naphthenic content may allow higher reactor temperatures to be used.

Once the feed is introduced and there is no evidence of temperature instability, reactor temperatures may be raised slowly, about 25°F/hr. There are various plateaus at which further increases in the reactor temperature are held up until moisture and sulfur in the recycle gas come down to specified levels. Within these limitations, the reactor temperature is raised to the level required to provide the desired product octane or aromatics yield.

G.   POISONS

During processing, upsets in the feed pretreatment or contaminants in the feed can poison the reforming catalyst. The effect of these poisons can be either temporary or permanent. Common temporary poisons include sulfur, nitrogen, and chloride. Permanent poisons typically encountered are lead and arsenic. Table III is a list of common reformer catalyst poisons.

1.   Sulfur

Sulfur poisoning usually results from an upset in or failure of the reformer feed pretreatment system. Although straight platinum reforming catalysts can tolerate a few parts per million of sulfur in the feed on a steady basis,

TABLE III
Common Reformer Catalyst Poisons

Temporary poisons	Permanent poisons
Sulfur	Lead
Nitrogen	Arsenic
Chloride	

bimetallic reforming catalysts generally require less than 1 ppm and some require less than 0.2 ppm. Sulfur poisoning affects the hydrogenation–dehydrogenation function of the catalyst. Indications of sulfur poisoning in a unit are loss of C_5+ and H_2 yields; loss of activity (a need to increase the reactor inlet temperature rapidly to maintain product octane); reduction in the magnitude of the temperature drop across the unit, especially in the first reactor; and the presence of H_2S in the recycle gas.

The activity loss due to sulfur poisoning can be recovered by removing the sulfur from the feed and continuing to process the feed. Unfortunately, when sulfur poisoning occurs, it is not always recognized as such. Consequently the refiner rapidly increases the reactor temperature to offset the loss of activity. The reactor temperature increase produces a more rapid coke deposition. This coke deposition results in a shortened cycle length even if the source of the sulfur is removed at a later time. When sulfur poisoning occurs and is correctly identified, a good practice is to reduce the reactor inlet temperature to about 900°F. This reduction in operating severity avoids additional, unnecessary coke deposition. The source or cause of the sulfur contamination should be located and corrected quickly to minimize the length of time the unit is run at reduced severity. After the feed sulfur specification is once again achieved, the unit should be kept at the reduced severity until the quantity of H_2S in the recycle gas drops to a few parts per million. At that point, the unit can be returned to the desired operating severity with little, if any, loss of cycle length or activity.

Once the source of sulfur contamination has been eliminated, a quicker way to reduce the sulfur on the reforming catalyst than low-severity feed processing is to strip with hydrogen. Feed is cut out of the unit, and reactor temperatures are raised to the 950–975°F range. Hydrogen is swept through the catalyst beds for a period of time, usually not exceeding 24 hr. The H_2S in the off-gas is monitored to determine when to end the hydrogen sweep.

2. Nitrogen

Nitrogen poisoning is also associated with an upset in or failure of the feed pretreatment. Nitrogen poisoning affects the acid function of the catalyst by forming ammonia and neutralizing catalyst acidity and by forming ammonium chloride and stripping chloride off the catalyst. The frequency of nitrogen poisoning is less than that of sulfur poisoning, since feeds are usually low in nitrogen content even before feed pretreatment. Notable exceptions to this are naphthas from West Coast crudes and from synthetic crudes such as shale oil. Most catalysts require feed nitrogen levels of less than 1 ppm to less than 0.5 ppm.

Indicators of nitrogen poisoning are a shift in catalyst selectivity to lower quantities of the iso form of the hydrocarbons as the result of a reduction in isomerization reactions and increased difficulty in maintaining product octane but without much loss in the temperature drop across the unit. If this condition continues for a sufficient period of time, ammonium chloride will precipitate in the feed–effluent heat exchangers with a resultant loss in efficiency. As with sulfur poisoning, if the refiner attempts to offset the activity loss by increasing the reactor temperature, the rate of coke deposition will be increased. After the source of nitrogen poisoning is eliminated, the nitrogen in the reformer is removed by continued feed processing. Because of the loss of chloride, it will be necessary to replenish the removed chloride to return the catalyst to the desired balance between the hydrogenation–dehydrogenation function and the acid function.

3. Chloride

Although it is desirable to maintain the chloride level on the catalyst at a predetermined value, occasionally chloride enters with the crude oil from certain oil well cleaning or enhanced recovery techniques. When this happens, the chloride level on the reforming catalyst increases and unbalances the dual functions of the catalyst. Excessive chloride on the catalyst shifts yield selectivity and frequently results in excessive hydrocracking.

Evidence of a high chloride level is a loss in hydrogen yield, increase in recycle gas gravity, loss in C_5+ yield, reduction in the temperature drop in the last reactor (sometimes even a temperature rise in cases of excessive hydrocracking), and an increased level of HCl in the recycle gas stream. The effects of high levels of chloride in the feed can be partially offset by adding water or alcohol to the feed to wash off the excess chloride. The heat exchangers can be water-washed to remove the ammonium chloride deposits.

4. Lead

Lead poisoning usually results from the feed becoming contaminated either from being transported in a tanker or barge that previously contained leaded gasoline or from rerunning leaded gasoline from refinery slop. When the quantity of lead in the feed exceeds the ability of the pretreater to remove it, lead enters the reformer and permanently deactivates the catalyst by interfering with the hydrogenation–dehydrogenation function.

Lead poisoning is characterized by a loss of activity in the first reactor, most easily detected by observing a decrease in the temperature drop across the reactor. Although eliminating the cause of the lead poisoning limits further deactivation, the activity that has been lost is not recoverable. Lead tends to build up on the first catalyst it contacts before moving deeper into the catalyst bed or into downstream reactors. Replacement of the catalyst in the first reactor is frequently required in cases of lead poisoning. A guideline for acceptable feedstock lead levels for bimetallic reforming catalysts is less than 10 ppb.

5. *Arsenic*

Arsenic, contained in some crudes, acts like lead in poisoning reforming catalysts. The sensitivity of reforming catalysts to arsenic poisoning is even greater than to lead poisoning. A guideline to acceptable feedstock quality for arsenic content is less than 2 ppb.

H. COKING DEACTIVATION

As the run progresses, even if no poisoning problems occur, the catalyst eventually loses activity as a result of coke deposition. The rate of coke deposition is a function of the feedstock quality, the operating severity of the unit, the unit pressure, the LHSV, H_2:HC molar ratio. As the carbon level on the catalyst increases, reactor inlet temperatures have to be raised to offset the loss in activity. In semiregenerative units, carbon characteristically deposits lowest in the first reactor and highest in the last reactor. The level of carbon continues to build until the end of the cycle is signaled by the unit's heaters inability to raise the reactor inlet temperature further at a constant feed rate. Because of changes in the selectivity of the catalyst as the quantity of coke increases, some refiners find economic justification in ending a cycle before reaching the unit's heater limitation.

For the bimetallic reforming catalyst in a semiregenerative unit, carbon levels in the last reactor of as much as 20–25 wt% have been reported at the end of the cycle [7]. Although cycle lengths are highly variable and dependent upon most of the unit's operating conditions and feedstock quality, most refiners with semiregenerative units attempt to achieve minimum cycle lengths of 6 months to a year. In cyclic units and in moving bed units, the carbon levels on the catalyst are held to much lower values.

I. REGENERATION

1. *Purpose*

Regeneration is the process of restoring a reforming catalyst to its original activity by carefully removing accumulated coke deposits. During the regeneration process, measures can be taken to overcome the effects of temporary catalyst poisons if the unit was shut down before their deactivating effects were eliminated.

2.   Procedure

A typical regeneration procedure is a multistep oxidation process. After feed is removed from the unit, the catalyst is swept with recycle gas for a period of time to remove heavy hydrocarbons left in the unit. The heaters, reactors, and recycle gas system are isolated from the remainder of the unit. The hydrogen atmosphere is replaced with a nitrogen atmosphere. At 700 to 800°F, a small quantity of oxygen is admitted to the system to initiate burning of the coke. Temperatures within the reactors are carefully monitored during the burn to avoid excessive temperatures which could damage the catalyst. The carbon is removed in a series of steps in which either the temperature or the oxygen concentration of the regeneration gas is increased until there is no further evidence of coke combustion.

Although most refiners introduce oxygen into the first reactor of a semiregenerative unit and continue the regeneration sequence to the last reactor, some refiners introduce oxygen to both the first and last reactors at the same time. This is done as a time-saving step, since the large quantity of coke in the last reactor takes the longest time to remove. However, the oxygen content of the gas entering the last reactor must be carefully monitored. Upon completion of the combustion of coke in the reactor ahead of the last reactor, the oxygen no longer being consumed enters the last reactor, which when combined with the oxygen content already present can result in excessively high temperatures.

When the unit must be shut down for maintenance, maintenance is done either before or after a proof burn which is a burn conducted at a high temperature and oxygen content. After maintenance or the proof burn, the unit is ready for reduction, presulfiding, and the introduction of feed.

Between regeneration and reduction, the refiner may apply other proprietary procedures, as recommended by either the process licensor or catalyst vendor, to ensure good precious metals dispersion on the catalyst.

The regeneration procedure for cyclic and moving bed units also follows the process licensor's recommendations.

3.   Source of Oxygen for Regeneration

In semiregenerative units, the oxygen used in regeneration can come from air or from liquefied oxygen. Regeneration at low pressure, e.g., 100 psig, with air as the source of oxygen is the most common method.

By using pure oxygen in a nitrogen atmosphere, a refiner can regenerate the catalyst at a higher pressure. This reduces the time required for the regeneration procedure, although the hazards associated with using pure oxygen must be carefully considered.

4. Offsite Regeneration

As a result of the success in other catalytic processes, some refiners are investigating having their reforming catalysts regenerated by merchant catalyst regeneration services. In this case, the coke-laden catalyst is removed from the reactors and shipped to the merchant regenerator's plant where the carbon is removed by oxidation. The regenerated catalyst is then returned to the refiner for reloading. Merchant regeneration may offer some advantages because of the generally better temperature control afforded as compared with *in situ* regeneration.

J. TOTAL LIFE

If a reforming catalyst is protected from permanent poisons, it can be returned to its original activity by careful regeneration. With proper care, in semiregenerative units, catalysts have been used for at least 10 cycles before being replaced. Bimetallic reforming catalysts have lasted for over 10 yr of operation or 800 bbl/lb in some reformers [8]. In cyclic units, it is not uncommon for catalysts to be subjected to several hundred regenerations before being replaced.

When it has been determined that a reforming catalyst can no longer have its activity restored by regeneration, the catalyst is replaced. The spent catalyst is usually sent to a metals reclaimer for recovery of the platinum and, if present, other promoter metals, especially rhenium. The recovered precious metals are normally returned to the catalyst manufacturer for incorporation into the next production of fresh catalyst for the refiner.

V. Operating Variables

A. REACTOR INLET TEMPERATURE

The operating variable used most by unit operators is the reactor inlet temperature. Although the size of the unit and the capacity of the pumps and compressors are fixed at the design stage, the reactor inlet temperature can be varied to achieve and maintain the desired product properties. Several temperature profiles are commonly used, as diagrammed in Fig. 6.

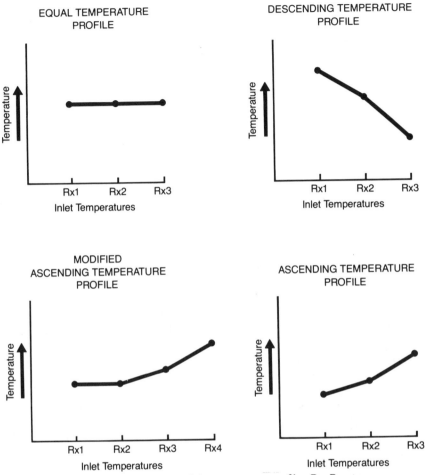

Fig. 6. Example of reactor inlet temperature profiles. Rx, Reactor.

Probably the most common profile is one in which all the reactor inlet temperatures are the same. As the coke on the catalyst builds up, the inlet temperature of each reactor is increased equally. This procedure is followed until the limit of the unit's ability to achieve higher temperatures is reached.

Another profile used has descending reactor inlet temperatures. In this case, the lead reactor temperature is higher than that in the next reactor in line and so on to the last reactor, which has the lowest reactor inlet temperature. As with the even temperature profile, the individual reactor inlet temperatures are increased to compensate for catalyst deactivation while maintaining the descending profile. The rationale for the descending profile is to keep the last reactor temperature lowest to minimize coke deposition, since with an even temperature profile carbon deposition is highest in the last reactor. Unfortunately, even with a 5–10°F difference in inlet temperature between reactors, the average bed temperature of the last reactor is still higher than the average bed temperature of the lead reactor. Thus, although possibly helping to prolong run length, the average bed temperature, which affects carbon deposition, is still higher in the last reactor and so the carbon deposition is higher also.

Another profile encountered has the first two of four reactors held at start-of-run conditions for most of the cycle. To maintain product quality the reactor inlet temperature of the third and fourth reactors is increased as the catalyst deactivates. This profile is typical of units with a split recycle. In these units a small quantity of recycle gas is fed to the first two reactors where dehydrogenation reactions normally occur. Since the average bed temperature in the first two reactors is relatively low, and since the reactions occurring are not the major contributors to coking, a high $H_2:HC$ molar ratio is not required to minimize coke formation. Recycle gas from the first two reactors combines with additional recycle gas injected ahead of the third reactor. This provides a higher $H_2:HC$ molar ratio in reactors in which the major coke-producing reactions occur.

Even in units without a split recycle, occasionally an ascending reactor inlet temperature profile is tried. Based on the consideration of long cycle length, the ascending profile appears to be the least desirable, since it aggravates the already high average bed temperature in the last reactor.

B.   FEEDSTOCK END POINT

The refiner may have some flexibility in the choice of feedstock end boiling point. The end boiling point of the feed is somewhat set by the purpose for which it is reformed. Since the reformate has an end boiling point about 15–20°F higher than that of the feed, the motor fuel reformer feed is usually limited to a maximum end point of 400°F to meet gasoline boiling specifications. Feeds to BTX reformers generally have a lower end boiling point of about 300°F.

In the case of motor fuels reformers in which the feed end boiling point may exceed 400°F, the portion of the feed that boils above 400°F greatly increases coking of the catalyst. Because of the rapid coking tendency, most refiners do not process feed with an end boiling point exceeding 400°F.

C. WATER–CHLORIDE BALANCE

An operating variable that has a major impact on the performance of the reforming catalyst is the moisture level in the unit. After pretreating, the naphtha entering the reformer typically contains 3–5 ppm by weight water. In the reformer the water in the feed vaporizes and interacts with the chloride on the reforming catalyst and will strip the chloride from the catalyst if given enough time. If the water level in the feed is very high, e.g., 50 ppm, it can remove chloride from the catalyst quickly and upset the dual function balance of the catalyst. To monitor the quantity of moisture in the unit, refiners use recycle gas moisture analyzers. With information from these moisture analyzers, refiners can add chloride to the reformer feed at levels prescribed by the process licensor or the catalyst vendor to maintain the proper metal–acid function balance.

Some process licensors and catalyst vendors feel that their catalysts work best in an essentially moisture-free atmosphere with 1 ppm by volume water or less in the recycle gas. Where great efforts are made to avoid contacting these catalysts with water, chloride addition is not required.

Because moisture displaces chloride from the catalyst, many refiners tended to slug chloride into the unit when it neared its end-of-run temperatures. In some cases this increased the product octane and allowed the cycle to be stretched for several weeks before regeneration was required. In some cases the slug of chloride was too much and it upset the operation. In general, optimum performance is achieved by adding small quantities of chloride continuously during the run to establish a water–chloride balance that maintains the desired chloride level on the catalyst. These balances are usually specific for each catalyst type and considered proprietary information by the process licensor or catalyst vendor.

VI. Future Reforming Growth

Although catalytic reforming is a firmly established refinery process that will continue to be used, its additional growth in capacity is tied to gasoline consumption, automobile engine design, and petrochemicals demands. The switch to lead-free gasoline that started in the mid-1970s, necessitated by government regulation of automobile exhaust emissions, strained refiners' ability to provide enough high-octane blending components for their gasoline pools. Reformers were run to higher octane levels and in some cases expanded in anticipation of future demands.

Currently, however, because of a combination of the economic pressures of high gasoline costs and automobile manufacturers' improvements in automobile mileage, gasoline consumption has dropped considerably from previously projected values. Automobile makers are committed to manufacture even smaller, more fuel-efficient automobiles in the future. Consequently, it appears that the need for expansion of reformer capacity will be limited for at least the next several years.

References

1. W. L. Nelson, "Petroleum Refinery Engineering," 4th edition, p. 810. McGraw–Hill, New York (1958).
2. F. G. Ciapetta, and D. N. Wallace, *Cat. Rev.* **5**(1), 67–158 (1971).
3. L. R. Aalund, *Oil Gas J.* **79**(13), 63–65 (1981).
4. B. C. Gates, J. R. Katzer, and G. C. A. Schuit, "Chemistry of Catalytic Processes." McGraw–Hill, New York (1979).
5. W. H. Hatch, S. J. Cohen, and R. Diener, Modern Catalytic Reformer Designs Help Reduce Cost of Low-Lead Gasoline. Presented at NPRA Annual Meeting, AM–73–33 (1973).
6. NPRA Q&A Session on Refining and Petrochemical Technology, pp. 28–29. Petroleum Publishing, Tulsa (1976).
7. NPRA Q&A Session on Refining and Petrochemical Technology, p. 90. Farrar and Associates, Tulsa (1977).
8. NPRA Q&A Session on Refining and Petrochemical Technology, pp. 23–24. Petroleum Publishing, Tulsa (1976).

CHAPTER 6

Catalysis of the Phillips Petroleum Company Polyethylene Process

J. PAUL HOGAN

Research Center
Phillips Petroleum Company
Bartlesville, Oklahoma

I. Discovery and Development

The Phillips polyolefins process, which by 1980 accounted for over 5 billion lb annually of polyethylene produced and sold worldwide, grew out of a discovery made less than three decades earlier at the Phillips research laboratories. This discovery (made by the author and his associate R. L. Banks) that ethylene could be converted to solid polymers over a chromium oxide–silica–alumina catalyst [1–3] had as its genesis an earlier discovery at Phillips, by Bailey and Reid, that ethylene could be converted to liquid polymers over a certain nickel oxide–silica–alumina catalyst [4].

The initial polymerization system involved a fixed bed of catalyst, but within 2 yr a more promising system involving solution polymerization in a stirred reactor was developed in the laboratory. Pilot plant studies in the research and development (R&D) department were begun within a year or two of the initial discovery. By about 1954, these studies had progressed far enough to permit commercial process design to begin. During this period, the pilot plant studies had progressed from batch operation, as done in the laboratory, to an integral continuous-flow system capable of producing a pelletized product [5].

Process design [6–17], by the R&D department, of commercial plants to manufacture high-purity ethylene and to convert the ethylene to high-density polymers by the Phillips process was completed rapidly. The engineering department of Phillips provided the mechanical design.

The applications research and market development of the new high-density polyethylene was begun in the R&D department as soon as laboratory samples were available. This was pioneering work, because the new Phillips resins were markedly different from the softer, lower modulus polyethylenes of the high-pressure process [18–20]. A market development and sales service laboratory was organized by Phillips Chemical Company in 1955 [21].

Characterization of the structure of the new polyethylenes by various physical and chemical methods showed that they were unique among the olefin polymers available at that time. They were shown to possess a high degree of structural simplicity and homogeneity and to consist of unbranched polymethylene chains terminating at one end in a vinyl group and at the other end in a methyl group [22].

In April 1955, on the recommendation of the technical and market evaluation groups, Phillips management approved the building of a commercial complex consisting of a 75-million-lb/yr polyethylene plant and a 180-million-lb/yr high-purity ethylene plant.

It was concluded that no one manufacturer could satisfy the market potential of the linear olefin polymers, and Phillips decided to license the process. In 1955 and 1956, nine companies in seven countries became licensees. Each company was supplied with complete technical information and plant designs and was furnished polymer for evaluation. To supply Phillips and its licensees with market development polymer, engineering data, and training in plant operation, a 1000-lb/day semiworks unit was put in operation in Bartlesville in 1955. This unit was scaled down from the commercial design, making it invaluable in the successful start-ups of the first commercial plants.

The Phillips commercial plant began operations at Pasadena, Texas, in December 1956. Licensee plants were not far behind in their start-ups. The first commercial grades of polyethylene produced by the Phillips process were limited to homopolymers of various melt indexes. Ethylene-1-butene copolymers [20] were introduced in 1958, and soon thereafter other polymer parameters were utilized to extend the choice of polymer types.

Commercial production by the Phillips process was entirely by the solution process [23] for the first 4 yr. Meanwhile, laboratory discoveries [24] in the early 1950s led to a more streamlined process, referred to as the particle form process. In this system a hydrocarbon that was a poor solvent was used as a liquid diluent in which to suspend the catalyst. The polymer formed as discrete particles, starting with each catalyst particle. This batch particle form slurry system was developed through extensive pilot plant work into a continuous commercial system which went on stream at the Phillips plant at Pasadena, Texas, in 1961 [25]. This system, also used by the licensees, soon accounted for more than half the total production by the Phillips process and brought about a further increase in the product spectrum.

The Phillips process capability was later extended to include ethylene-l-hexene copolymers and linear low-density polyethylenes.

II. The Catalyst System

The catalyst is the heart of the Phillips process, distinguishing it from the high-pressure process both in the conditions of polymerization and the properties of the products. A heterogeneous catalyst having a high surface area is used, and only moderate conditions of temperature and pressure are required for polymerization.

The two important features of the catalyst are chromium in the Cr(VI) state and a powdered substrate of silica or silica–alumina, although other substrates such as zirconia, thoria, or germania [1, 26] may be used. The concentration of Cr(VI) in the catalyst can vary from a few one-hundredths of a percent to several percent, but typically the concentration of total chromium is 0.5–1%. Various compounds of chromium, such as oxides, chromates, dichromates, halides, sulfates, nitrate, and organic compounds of chromium may be used as a source of chromium ion. The chromium compound is normally distributed on the support from aqueous solution by simple impregnation, although nonaqueous solutions, dry grinding, and various other methods have been employed. In fact, it was found at the Phillips laboratories that metallic chromium powder could be mixed with the silica or other support to obtain a typically active catalyst following heating in air [27].

After the chromium compound and substrate are combined and dried sufficiently to give a free flowing powder, the catalyst is activated by heating in air to a high temperature. This can be done in a fluidized bed at temperatures of 500 to 1000°C, dry air being used to produce fluidization [28]. Following a programmed heat-up period, the temperature is held constant for 5 to 10 hr while fluidization is continued. The catalyst is then allowed to cool, flushed with nitrogen to remove the air, and stored under dry nitrogen pressure until used.

During activation, adsorbed water is completely removed from the catalyst, and a large percentage of the silanol groups are removed in the form of water. During this dehydration process the chromium ions largely combine with the silica substrate in the Cr(VI) state, regardless of their starting valence. It has been shown that a surface silyl chromate forms by the reaction of Cr(VI) with surface silanol groups [28]. A small amount of dichromate may also form.

$$
\begin{array}{c}
\text{H} \quad\quad \text{H} \\
\text{O} \quad\quad \text{O} \\
| \quad\quad\quad | \\
-\text{Si}-\text{O}-\text{Si}- \ + 2\text{CrO}_3 \ \rightleftharpoons
\end{array}
\quad
\begin{array}{c}
\quad \text{O} \quad\quad\quad \text{O} \\
\quad \| \quad\quad\quad \| \\
\text{O}=\text{Cr}-\text{O}-\text{Cr}=\text{O} \\
\quad | \quad\quad\quad | \\
\quad \text{O} \quad\quad\quad \text{O} \\
\quad | \quad\quad\quad | \\
-\text{Si}-\text{O}-\text{Si}- \ + \ \text{H}_2\text{O}
\end{array}
\qquad (1)
$$

$$
\begin{array}{c}
\text{H} \quad\quad \text{H} \\
\text{O} \quad\quad \text{O} \\
| \quad\quad\quad | \\
-\text{Si}-\text{O}-\text{Si}- \ + \text{CrO}_3 \ \rightleftharpoons
\end{array}
\quad
\begin{array}{c}
\text{O} \quad\quad\quad \text{O} \\
\ \diagdown \quad\quad \diagup \\
\quad\quad \text{Cr} \\
\ \diagup \quad\quad \diagdown \\
\text{O} \quad\quad\quad \text{O} \\
| \quad\quad\quad | \\
-\text{Si}-\text{O}-\text{Si}- \ + \ \text{H}_2\text{O}
\end{array}
\qquad (2)
$$

Both Eqs. (1) and (2) are reversible, and it is necessary to activate the catalyst in relatively dry air, especially when the temperature is above 800°C, to prevent decomposition of CrO_3 on the surface to Cr_2O_3. In the absence of a surface structure, CrO_3 readily decomposes to Cr_2O_3 even in air at 500°C [28]. Under the best activation conditions, greater than 75% of the chromium remains in the Cr(VI) state on the surface of a silica support.

Although the principal oxidation state of Cr on the catalyst after activation is Cr(VI), under some conditions relatively high concentrations of Cr(III) may be present. This is illustrated in Table I, which presents data on chromium oxide–silica catalysts containing varying amounts of total Cr that have been activated in dry air at 540 and 800°C and in wet air at 540°C. Activation in each case was 5 hr in a fluidized bed at an air space velocity of 500 volumes of air per hour per volume of catalyst.

The Cr on the catalyst was largely Cr(VI) at a 540°C activation temperature over the range of 1 to 5% total Cr loading (Table I). However, at higher loading or at a higher activation temperature, the percentage of Cr(VI) dropped sharply. Even at 540°C Cr(VI) decreased sharply when wet air was used as the activation gas. High Cr loading caused conversion of CrO_3 to

TABLE I

Chromium Oxide–Silica Catalysts Activated in Air

Activation conditions		Catalyst after activation	
Temperature (°C)	Air	Total (wt%)	Cr(Vi) (% of total Cr)[a]
540	Dry	1.0	93
540	Dry	2.5	94
540	Dry	4.9	94
540	Dry	11.8	57
540	Wet[b]	2.5	81
540	Wet[b]	4.9	33
800	Dry	1.0	85
800	Dry	2.5	87
800	Dry	5.0	65

[a] Determined independently by solubility and by reaction with Fe^{2+}.
[b] Saturated with water vapor at 38°C.

Cr_2O_3 because of CrO_3 clumping. The presence of water in the activation air caused a reversal of Eqs. (1) and (2), resulting in the conversion of free (unstabilized) CrO_3 on the surface to Cr_2O_3: $2CrO_3 \rightarrow Cr_2O_3 + \frac{3}{2}O_2$. Because Cr_2O_3 on the catalyst surface is relatively inert, its presence is of no consequence provided there is enough Cr(VI) present to provide a sufficient concentration of active sites for good catalyst activity.

The physical structure of the silica gel used for catalyst manufacture was found to be of extreme importance. Surface area was found to influence activity, and pore size was found to affect both catalyst activity and the molecular weight of the polymer [29]. Consequently, silica gels with increased pore sizes, not available commercially, were developed in the Phillips R&D department [30, 31]. Catalysts ranging in average pore diameter from 50 to above 200 Å are now used in Phillips polyethylene plants. Other catalyst and process developments further extended the range of molecular weights available in the particle form process [32–34].

In some cases catalyst adjuvants may be used to modify polymer properties. Diethyl zinc, triethyl boron, and combinations of these and other zinc and boron alkyls can be used with a chromium–silica catalyst to broaden the molecular weight distribution [35]. Various fluorine compounds may be added to the catalyst before activation to increase activity or modify polymer properties [36]. Pretreatment of the catalyst at elevated temperature with carbon monoxide [37] may be used to modify catalyst activity or polymer properties.

III. Polymerization

A. GENERAL PRINCIPLES

The polymerization of ethylene with Phillips catalyst systems may be done over a relatively broad temperature span ranging from ambient to about 200°C. However, the temperature range used commercially is approximately 65–180°C. In general, the higher the reaction temperature, the higher the melt index of the polyethlene.

Ethylene may be polymerized from a hydrocarbon solution or as a compressed gas. When a hydrocarbon that is a good solvent for polyethylene is used and the temperature is sufficiently high, the polyethylene remains in solution (solution process). If a poor solvent is used and the temperature is sufficiently low, the polyethylene forms as granules, with the catalyst as a nucleus (the particle form process). In the gas phase process the polymer is also granular.

The rate of ethylene polymerization depends on catalyst activity, ethylene concentration in the reaction medium, reaction temperature, and polymerization time (catalyst age). When the polymerization temperature is low and the catalyst has not been pretreated at an elevated temperature with CO or a hydrocarbon, an induction period of a few minutes occurs before the catalyst sites become activated for rapid polymerization. This activation involves an oxidation–reduction step which will be discussed in Section VI.

The catalyst is quite sensitive to catalyst poisons, and feed impurities such as water, oxygen, and many covalent compounds of oxygen, nitrogen, sulfur, and halogens are catalyst poisons. Most of these compounds are polar and may act as catalyst poisons by binding tightly at the polymerization site, thus preventing the adsorption of olefin monomer and the initial activation of the site, or preventing the propagation of a growing chain if introduced after the reaction is underway. Poisoning is generally of a temporary nature, since activity may recover if the addition of poison to the reactor is stopped.

Copolymerization of ethylene with α-olefins over the Phillips catalyst modifies the density of the polymer and improves such properties as flexibility and stress crack resistance. Incorporation of the α-olefin results in the formation of short branches which interrupt crystallization and cause the formation of a softer, less dense polymer.

B. SOLUTION PROCESS

The solution process for producing polyethylene is illustrated in Fig. 1. This is a continuous-flow system having five main steps: (1) conversion of

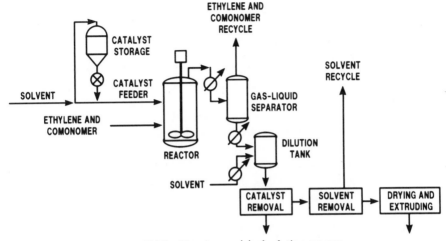

Fig. 1. Phillips Petroleum original solution process.

ethylene or ethylene plus α-olefin to the polymer in a stirred reactor; (2) separation of unreacted monomer from the solvent–polymer solution; (3) removal of catalyst from the polymer solution; (4) separation of the polymer from the solvent; (5) extrusion of the polymer, pelletizing, and packaging.

The solvent used in the process must be volatile enough for easy separation from the polymer but not so low-boiling as to cause excessive pressure in the first three steps of the process. A paraffin and/or cycloparaffin hydrocarbon mixture that is free of catalyst poisons, low in aromatics, and boils in the 65–95°C range is ideal. The solvent is prepared for the process by degassing to remove air and then passing over a desiccant such as silica or alumina.

The ethylene monomer is dried in desiccant beds and continuously monitored for purity. Comonomer, such as 1-butene, 1-hexene, or other α-olefins, is stored in vessels from which it is transferred through degassing and drying equipment and finally charged to the reactors as a liquid by means of pumps and metering devices.

The catalyst powder is stored in bins under a dry atmosphere after activation and before use is flushed with nitrogen to remove entrained air. The catalyst may be charged to the reactor by various techniques such as dry auger–feeders, pumping of dilute hydrocarbon slurry with a positive displacement pump, or feeding as a "mud" (a settled slurry in hydrocarbon).

A number of reactor designs may be used in solution polymerization with the Phillips catalyst [1, 8]. The reaction temperature ranges up to 180°C. The reactor is operated liquid-full at pressures in the range 20–30 atm, an operating pressure being chosen that ensures that the monomer–solvent mixture remains in the liquid phase at the operating temperature in use. The reactor design incorporates the following features: (1) good mixing to ensure that the catalyst is suspended and the polymer solution makes intimate contact with heat transfer surfaces; (2) heat removal (1450 Btu/lb of ethylene polymerized) at surfaces that are not overcooled, thus avoiding plating out of the polymer; (3) accurate temperature control for control of the polymer melt index.

The operation of the reactor is continuous. The feed rates of the catalyst, the monomer, and the solvent are balanced to provide the desired rate of polymer production, the desired monomer concentration (3–10 wt% of reactor contents), and the optimum polymer concentration, which may vary from 6 to 15% depending on the melt index of the polymer being produced. However, polymer concentrations as high as 25% have been used commercially in reactors of special design. In the production of copolymers, an α-olefin is charged continuously to the reactor as a separate stream at a rate that will maintain the proper concentration in the reactor.

An effluent stream, having the same steady-state composition as the reactor contents, is continuously withdrawn automatically to maintain a constant reactor pressure. The effluent is first passed through flash or fractionation steps to remove unreacted ethylene monomer which is recycled to the reactor. Ethylene conversion per pass is a function of the steady state polymer and ethylene concentrations in the reactor. For example, in a reactor operating at 5 wt% ethylene concentration and 15 wt% polymer concentration, the conversion per pass is always 75%. Since the unreacted ethylene is recycled, the overall conversion approaches 100%.

After the ethylene is removed from the effluent polymer solution, the solution is centrifuged or filtered in special equipment to remove catalyst residues [16]. Catalyst removal is not necessary in some cases when catalyst productivity is above about 2000 lb of polymer per pound of catalyst.

The polymer is typically precipitated from solution by intimately mixing the polymer solution with water, followed by stripping out of the remaining solvent with steam [7, 9, 14, 38, 39]. In more recently designed processes, only flash devolatilization is used. In each case the solvent is recycled to the reactor after the removal of water and polymer impurities [7].

After the polymer is separated from the solvent, it is dried to remove residual water or solvent and is conveyed to bins where it is fed to single- or twin-screw extruders operating at about 200–250°C. Additives such as antioxidants and stabilizers may be fed continuously to the stream of "fluff" going to the extruder. The discharge end of the extruder is equipped with a screen pack and a multiple-hole die. The polymer extruding from the die face is pelletized directly by means of an underwater die face chopper. The pelletized polymer is bagged, boxed, or fed directly to hopper cars for shipment to the user.

C.   PARTICLE FORM LOOP REACTOR PROCESS

The Phillips particle form (slurry) loop reactor process is depicted in Fig. 2 [34]. The main steps in this process are (1) polymerization, (2) separation of monomer and diluent from the polymer granules, and (3) pelletization (if desired). Catalyst removal is not practiced because the catalyst productivity is high, ranging from 3000 to greater than 10,000 lb of polymer per pound of catalyst, and the catalyst residue is insignificant. Because the polymer forms as granules, each polymer granule originating with a tiny catalyst particle, the precipitation step used in the solution process is eliminated.

A low-boiling hydrocarbon that is a poor polyethylene solvent is used as a diluent for suspending the catalyst, dissolving the ethylene gas, and transferring the heat of reaction to the cooling surfaces of the reactor. A reactor temperature below that at which the polymer dissolves or swells in the diluent is used in polymerization. As the polymer forms in the catalyst pores, the original catalyst particle is bound together with polymer into a porous crumb of polymer containing fragmented catalyst particles throughout. Polymerization activity may increase with time over a period of an hour after initiation as new sites are created [40].

Like the solution process, the particle form process is continuous. Catalyst, monomer, and diluent are fed continuously to a reactor, and a polymer

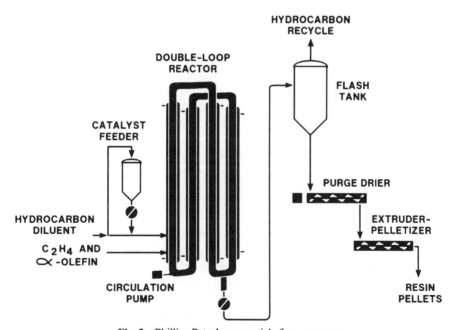

Fig. 2.   Phillips Petroleum particle form process.

slurry is continuously withdrawn. The reaction temperature is controlled by means of a coolant which is circulated through the reactor jacket or coils. The upper limit of reactor temperature varies with the diluent being used and other factors. Reactor pressure is maintained at a point sufficient to ensure that the system is liquid-full, thus maintaining all the ethylene feed in the liquid phase. The ethylene concentration is typically 2–6 wt% of the liquid phase, and the polymer concentration is 25–35%. Ethylene conversion per pass exceeds 95%.

Reactor design is critical in this process. A large amount of heat must be removed because of the high polymerization rates, and heat transfer surfaces must be kept clean. Temperature control must be precise for good melt index control. Initial development of this process in the laboratory and small pilot plant stages in the early and middle 1950s was done with stirred autoclaves. However, it was soon found that better efficiency and operability could be achieved through the use of a newly developed circulating loop reactor [41]. This concept was further developed, and a circulating loop reactor was put into commercial operation by Phillips in 1961. Subsequent engineering and process innovations have been combined to bring production capacity up to 13 lb of polymer per hour per cubic foot of reactor volume.

In the particle form process, the polymer melt index is controlled by varying the reactor temperature, by varying the composition or activation of the catalyst, or by the addition of hydrogen gas [42].

Effluent from the reactor is removed through a quiescent zone to permit settling of the polymer particles and removal of a minimum of diluent and ethylene from the reactor. The concentrated slurry is discharged into a flash tank where the ethylene and diluent vapors are withdrawn overhead for recycle and the polymer crumb drops into a conveyor–dryer. Remaining hydrocarbon is vaporized off the polymer crumb, and the dry crumb is then conveyed to storage, from which it is fed to extruders for pelletization and preparation for shipment. The polymer crumb may be sent directly to the processor without prior pelletization.

D. GAS PHASE POLYMERIZATION

The hydrocarbon diluent used in the particle form process is a convenient medium in which to suspend the catalyst and transfer the heat of reaction, but it is not essential. Experimental work at the Phillips laboratories in 1956 and 1957 showed that very high rates of polymerization could be obtained even in the absence of a diluent [8, 9, 43–45]. It was necessary only to keep the catalyst–polymer granules moving with respect to each other and with respect to reactor surfaces to ensure that the particles did not grow together into a mass or cling to the reactor walls. This was done in two different ways: (1) by fluidization with ethylene gas, and (2) by mechanical agitation such as with an anchor-type stirrer.

The gas phase process was further developed by a Phillips licensee, Badische Anilin- & Soda-Fabrik (BASF), in Germany in the 1960s [46]. Union Carbide commercialized a fluid bed gas phase system using a Cr catalyst a few years later [47]. The gas phase process has the possible advantage of eliminating the cost of supplying and recycling a diluent, but heat transfer rates are lower than in the particle form process, resulting in lower production rates per unit volume of reactor.

IV. Polymer Properties

Phillips-type ethylene homopolymer produced in the solution form process has a completely linear structure with no appreciable short-chain branching. Polyethylene produced in the particle form process is also linear as far as the best technique for measuring branching can detect. However, in certain fractional melt index particle form polymers, there is circumstantial evidence that a minimal amount of long-chain branching is present (see Section VI.B.3).

The density of Phillips-type polyethylene homopolymer varies from 0.965 for high-melt-index (low-molecular-weight) polymer to about 0.960 for a melt index of 0.3–0.5 and as low as 0.94 for ultrahigh-molecular-weight polymer. The density decrease with molecular weight increase is related to chain entanglement. Very long molecules become entangled enough to prevent maximum crystallization. Broad molecular weight distribution (MWD) polymers tend to be slightly higher in density than narrow MWD polymers because the short molecules can align with segments of long molecules to increase crystallization.

For density control in Phillips-type polymers of various melt indexes, an α-olefin comonomer such as 1-butene or 1-hexene is used. Copolymerization results in the formation of branches two carbon atoms shorter in length than the comonomer. For example, 1-hexene produces butyl branches. The addition of a few branches per 1000 carbon atoms to the backbone lowers the density sharply by disrupting crystallization. Figure 3 shows some typical relationships between (1) α-olefin in the feed and in the polymer, (2) α-olefin in the polymer and polymer density, and (3) polymer density and crystallinity.

The density range normally produced in the solution process is about 0.946–0.965. In the particle form process, densities typically range from

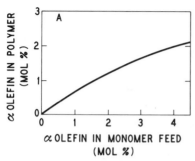 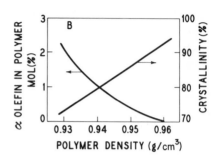

Fig. 3. (A) α-Olefin (1-C_6H_{12}) in feed versus in product. Loop reactor polymerization at 4 wt% C_2H_4. (B) Polymer density versus 1-C_6H_{12} content and crystallinity.

0.923 to 0.965. The effects of density changes at a constant melt index on physical properties are summarized in Table II. These data show that the polymer becomes less stiff and more ductile, and is much higher in environmental stress crack resistance (ESCR), as the density is lowered. The ESCR of these polymers can be further increased by the use of 1-hexene in place of 1-butene as comonomer, at the same density.

The melt index of polymers produced commercially by the Phillips process ranges from a value too low to measure up to a value of about 35. In the solution process, the range is typically 1 – 35; in the particle form process, the melt index ranges from 30 downward [33]. Extrahigh-molecular-weight grades produced in the particle form system do not have a measurable normal melt index, so a "10X" melt index is obtained at 10 times the normal extrusion pressure. This type of melt index (ASTM condition F) is often referred to as the high-load melt index (HLMI). The HLMI of extrahigh-molecular-weight polymers may vary from 1 to 10 or more, although a polymer having a HLMI of greater than 10 usually has a measurable regular melt index of 0.1 or more.

The physical properties of linear polyethylene that are the most sensitive to the melt index (or molecular weight) are correlated with the melt index in

TABLE II

Density-Dependent Properties of Ethylene-1-butene Copolymers (Melt Index 0.3)

	Density			
Property	0.96	0.95	0.94	0.925
Flexural modulus (psi $\times 10^{-3}$)	220	165	140	67
Tensile strength, 20 in./min (psi)	4400	3800	3300	1800
Elongation at break 20 in./min (%)	25	70	280	600
Bell ESCR, F-50 (hr)	30	150	600	>1000

TABLE III

Melt Index-Dependent Properties of Phillips Ethylene Homopolymers

Property	Melt index				
	0.2	0.9	1.5	3.5	5.0
Molecular weight, M_w	175,000	140,000	125,000	95,000	85,000
Tensile impact (ft lb/in.^{2})	100	64	59	41	30
Izod impact (ft lb/in. notch)	14	4	2	1.5	1.2
Elongation, 20 in./min (%)	30	25	20	15	12
Bell ESCR, F-50 (hr)	60	14	10	2	1
Brittleness temperature (°C)	−118	−118	−118	−101	−73

Table III. The ability to resist breaking under tensile impact declines with an increase in the melt index. The Izod impact strength of notched specimens declines more rapidly, indicating an increase in notch sensitivity as well as a decrease in impact strength. The ability to elongate (neck down) on pulling at a constant rate also declines appreciably over the range shown. Linear polyethylene remains flexible at low temperatures, even at a melt index of 5. The brittleness temperature becomes a function of the melt index only at higher melt indexes. The ESCR is very sensitive to the melt index. Ethylene homopolymers of high molecular weight (below 0.01 melt index) have ESCR values greater than 1000 hr.

The molecular weight distribution of both solution and particle form polymers is normally fairly broad but can be increased or decreased as needed by catalyst and process modifications. Typical weight average/number average molecular weight ratios (M_w/M_N) vary from as low as 3 for very narrow MWD resins to as high as 20 for very broad MWD resins. Narrow MWD polymers of 8 to 35 melt index are used in injection molding applications because of their resistance to warpage and high-impact strength. Broad MWD polymers with a fairly low melt index are used in applications requiring good ESCR and fast processing. Faster processing results from the increased shear response of broad MWD polymers. The following example illustrates this principle:

Polymer	M_w/M_N	MI	HLMI	HLMI/MI
A (narrow MWD)	6	0.3	18	60
B (broad MWD)	20	0.3	33	110

Even though both polymers have the same melt index (MI), polymer B has a much higher flow when the shear stress is increased 10-fold (HLMI).

Determination of the HLMI/MI ratio is a simple method of obtaining qualitative evaluation of the MWD of a polymer sample. However, other polymer structural properties such as long-chain branching can also affect the HLMI/MI ratio and must be taken into account.

V. Manufacturing Plants

The Phillips Petroleum Company plant for the manufacture of polyethylene is located at Pasadena, Texas, near Houston. Plant capacity as of the last quarter of 1982 was 1.4 billion lb/yr. Many other companies have licensed the Phillips process for polyethylene manufacture and have built plants. Some plants have changed ownership since construction. The following is a list of companies that have produced polyethylene under Phillips licenses in the 1970s: Allied Chemical Corporation, Chemplex Company, Gulf Oil Corporation, National Petro Chemicals Corporation, Soltex Polymer Corporation, Union Carbide Corporation (United States); BP Chemicals, Ltd. (United Kingdom); Manolene (France); Solvay & Cie (Italy); Rheinsche Olefinwerke GmbH. (West Germany); Calatrava (Spain); Polyolefins, N.V. (Belgium); Hemijska Industrija Pancevo (Yugoslavia); Showa Yuka, K.K. (Japan); Saga Petrokjemi a.s. & CO (Norway); and Eletrotena Industrias Plasticas S/A (Brazil). Canadian Industries Ltd., Canada, licensed the process in 1981.

VI. Polymer Structure and Polymerization Mechanism

A. STRUCTURAL BASIS OF MECHANISMS

A great deal of information about the molecular structure of various types of polyethylene has been obtained, principally by physical methods. On the other hand, even though much has been written about the mechanism of formation of high polymers of ethylene, at best these ideas have little solid information to substantiate them. The tools we have available for probing into the happenings at a catalyst site give us mainly circumstantial evidence, and our theories must be based in part on this evidence. However, we can at least require that theories of catalysis and reaction mechanism be compatible with information on polymer structure, which is on fairly solid ground. Therefore, this section has been designed to look first at the information on polymer structure and then to suggest polymerization mechanisms that seem compatible with this information and with catalyst structure.

B.   POLYMER MOLECULAR STRUCTURE

1.   Unsaturation and Short-Chain Branching

The molecular structure of ethylene homopolymer made by the Phillips process is that of a very high-molecular-weight olefin:

$$CH_2=CH(CH_2CH_2)_nCH_3$$

As the molecular weight is increased, n increases from the 100s to the 100,000s. Some of the structural features can be identified by infrared spectroscopy [22]. Curve A in Fig. 4 is the infrared spectrum of a Phillips homopolymer. Curve B is the spectrum of an ethylene-1-butene copolymer of about the same molecular weight as sample A but containing 2 mol% 1-butene comonomer. These spectra were produced with a Perkin-Elmer Model 137 Infracord spectrophotometer. The two curves representing samples A and B have several features in common. The large absorbences (bands) at 3.5, 6.8, and 13.9 μm are produced by methylenes that make up

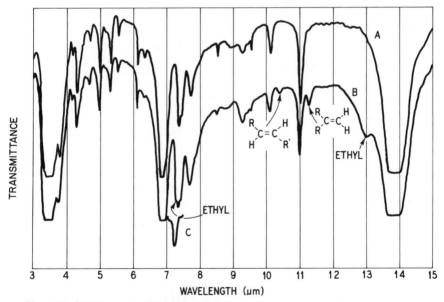

Fig. 4.   Infrared spectra of Phillips polyethylenes. (A) Particle form homopolymer of 1.7 melt index and 0.962 density. (B) Particle form C_2H_4 – 1-C_4H_8 copolymer of 1.7 melt index and 0.932 density. (C) Differential spectrum for A and B.

the polymer chains. The bands at 6.1, 10.1, and 11.0 μm are all owing to vinyl unsaturation.

The differences between curves A and B are apparent. A band at 10.35 μm in curve B indicates the presence of a small concentration of trans internal unsaturation. A band at 11.25 μm indicates the presence of a minor amount of branched vinyl. These two types of unsaturation, not apparent in sample A, are caused by the presence of 1-butene in the feed, and their significance will be discussed later in this section. The addition of ethyl branching from 1-butene incorporation in sample B has produced a band at about 12.95 μm and a shoulder at 7.25 μm that appears on the side of the larger 7.32 μm band. (The 7.25 μm absorbance is actually owing to the symmetric deformation of methyl and occurs regardless of the location of the methyl.) Although in this case the shoulder appears to be minor, it can be separated from the 7.32-μm peak by differential scanning, as shown by curve C in Fig. 4. In this case the 7.32-μm peak has been canceled out by homopolymer A placed in the reference beam of a Perkin-Elmer Model 21 unit [48].

Unfortunately, infrared spectroscopy does not readily distinguish branch length other than ethyl (at 12.95 μm) and methyl (at 8.7 μm). However, by means of new ^{13}C-NMR methods, the carbons in the branch and also alpha, beta, and gamma to the branch point along the backbone can be identified [49,50]. A ^{13}C-NMR spectrum for a Phillips-type ethylene-1-pentene copolymer is shown in Fig. 5. The chemical shifts in parts per million with respect to the tetramethylsilane internal standard change with changing branch length from methyl to hexyl, making possible the clear identification of branch length [50]. The number of branches of each length per 1000 carbon atoms in the molecule can be measured with precision by means of ^{13}C NMR [51].

2. Molecular Weight and Branching Distribution

The molecular weight distribution of Phillips polyethylene is polydisperse, as pointed out in Section IV. This is the result expected from a heterogeneous catalyst. Adjustments in molecular weight distribution by catalyst modifications involve catalyst changes that increase or decrease the heterogeneity of catalyst sites.

The distribution of branching with molecular weight in copolymers varies to some extent with polymerization conditions and possibly also with the α-olefin used as comonomer, but branching always increases in the lower molecular weight fractions. This is illustrated in Fig. 6, which is a plot of the branching concentration versus the weight average molecular weight (M_w) of fractions obtained by column elution fractionation of an ethylene-1-hex-

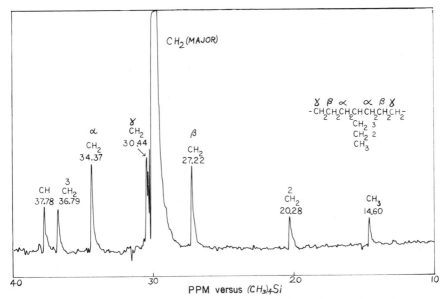

Fig. 5. ¹³C-NMR spectrum of Phillips-type $C_2H_4-1-C_5H_{10}$ copolymer. Chemical shifts in ppm are with respect to an internal tetramethylsilane standard.

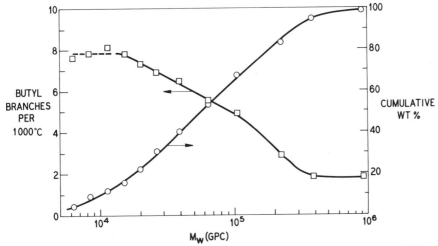

Fig. 6. Branching distribution in a particle form $C_2H_4-1-C_6H_{12}$ copolymer of 0.2 melt index and 0.940 density obtained by elution fractionation.

ene copolymer of 0.2 melt index and 0.94 density [52]. Also included is a plot showing the M_W of each fraction versus the accumulative weight percent of polymer fractionated at the midpoint of the fraction. The inverse relationship between branching concentration and molecular weight suggests a strong chain-terminating effect of α-olefin at sites that are usually active for copolymerization. The presence of branched vinyl unsaturation, which was especially prominent in the lower molecular weight fractions, is further evidence of termination by α-olefin, since this type of unsaturation does not occur in homopolymers.

3. Long-Chain Branching

Long-chain branching has been shown to be present in appreciable concentrations in many high-pressure polyethylenes. In this case, the free radical mechanism of polymerization accounts for long-chain branching. In Phillips-type polyethylene, it was pointed out in Section IV that there is evidence for the presence of a small amount of long-chain branching in particle form polymers. This evidence is based on the properties of the melt, the branching frequency being too low to be measurable by direct methods [53]. The example in Table IV compares two Phillips homopolymers. These two polymers are of almost identical solution viscosity and the same M_W, and yet the low-shear melt viscosity of the particle form polymer is 65% greater than that of the solution polymer. This difference in melt viscosity is in the direction opposite that expected from the differences in molecular weight distribution and must be accounted for by differences in molecular structure. Long-chain branching increases melt viscosity at a very low shear rate by causing constraints to flow. By contrast, in dilute solutions (or at high shear in the melt) long-chain branching may actually decrease viscosity by decreasing the size of a polymer coil of given molecular weight. Another effect of a small amount of long-chain branching is an increase in the

TABLE IV

Rheological Evidence of Long-Chain Branching in Particle Form Polymers

Polymerization method	Polymer properties			
	η^a	ηm^b	M_w, by GPC[c]	M_W/M_N
Solution	2.1	3.4	169,000	11
Particle form	2.0	5.6	169,000	7

[a] Inherent solution viscosity in tetralin (ASTM Method D1601).
[b] Melt viscosity in poises $\times 10^{-5}$, measured at very low shear (10^{-1}/sec).
[c] GPC, Gel permeation chromatography.

elasticity of the polymer melt, and this was demonstrated by studies with a Weissenburg rheogoniometer [53].

It is possible to account for long-chain formation in the Phillips particle form process without involving a free radical reaction. A discussion of this possibility will follow the presentation of a proposed mechanism for polymerization with the Phillips chromium oxide catalyst.

The formation of linear ethylene polymers over the Phillips catalyst precludes proposals that the polymerization mechanism is a free radical process. (In fact, free radical initiators poison the Phillips catalyst.) The absence of branching indicates that the mechanism is stereoregular.

C.   MECHANISM OF POLYMERIZATION CATALYSIS

It was pointed out in Section II that activation of the catalyst in dry air at an elevated temperature results in a major portion of the chromium ion forming Cr(VI) silyl chromate as a surface compound. This occurs regardless of the starting oxidation state. Also, a significant portion of the Cr may take the form of Cr_2O_3 (green) on the catalyst, particularly if the Cr concentration is above 1% or if much moisture is present at high temperatures. The Cr_2O_3 is inactive for polymerization.

A third oxidation state that has been identified in catalysts following activation is Cr(V). Much attention was given to studies of Cr(V) on this catalyst in the early 1960s (54–58). Some workers proposed that Cr(V) was the active species for polymerization. However, the highest concentrations of Cr(V) found at the Phillips Petroleum R&D laboratories occurred when catalysts that had been activated in dry air were exposed to a small amount of water vapor or when catalysts contained an excessive concentration of chromium oxides [59]. In both these cases the catalysts were found to be very low in activity for polymerization. Catalyst activity was not found to correlate with Cr(V) content at Phillips but did correlate with Cr(VI) concentration. The Cr(VI) content is normally at least 10 times as great as the Cr(V) content after catalyst activation, and the number of active sites measured [28] exceeds the Cr(V) content.

Although some of the theories on the structure of the active site that have appeared in the literature have pictured Cr(VI) or Cr(V) as the catalytically active species, these ideas are inconsistent with the chemistry of chromium. The polymerization of ethylene to high polymers over this catalyst occurs over a wide temperature range without indication of a significant change in reaction path. When ethylene first contacts an activated catalyst at 150°C, an oxidation–reduction reaction occurs within seconds, consistent with the ability of chromates to oxidize organic compounds. Even at 95°C, with a catalyst carefully cleaned by degassing at an elevated temperature, a color change from orange to blue was shown to occur, and polymerization began in less than a minute [28]. Most of the Cr was then present as Cr(II), and no Cr(VI) or Cr(V) could be detected [40, 59]. [Certainly, if polymerization is initiated at a low temperature or if the catalyst surface is partially blocked with poisons, some Cr(VI) or Cr(V) may remain, but this is beside the point.]

The Phillips catalyst can be pretreated with carbon monoxide to obtain a prereduced catalyst which initiates polymerization immediately without undergoing further reduction by ethylene. CO-reduced catalysts are quite low in oxidation state, being almost entirely in the Cr(II) state [28, 40, 60], and have quite high activity [37]. Thus the Cr(VI) can be concluded to be the precursor that forms the active site in a typical catalyst by an oxidation–reduction reaction with the feed. However, it was demonstrated at Phillips in the 1960s that active catalysts could be prepared that contained no Cr(VI) at any stage during preparation or activation. Some examples were $Cr(CO)_6$ on silica or silica–alumina [61], $Cr(CO)_6$ on alumina [62], bisarene chromium compounds on silica or silica–alumina [63], and Cr(II) chloride on silica [64].

In many catalyst systems the metal compound is the active catalyst and is supported or dispersed on a catalyst to increase the surface area. However, in the Phillips catalyst system, the silica or similar support is not merely a carrier but is essential for polymerization activity. The most efficient use of the Cr ion occurs when it is very dilutely distributed on the silica surface, but the most efficient use of the silica occurs when approximately 1% Cr is present. This is illustrated in Table V. Figure 7 is a graphical presentation of the data given in Table V.

The catalysts used to obtain the data shown in Table V were prepared by impregnating silica having medium-large pores with an aqueous CrO_3 solution, followed by drying and activation in a fluidized bed in dry air at 800°C. Polymerization testing was done in a batch reactor slurry system at constant temperature (slightly above 100°C) and 550 psig ethylene pressure. The run duration was 80 min after the onset of ethylene uptake. In these experiments the quantity of catalyst charged was adjusted in each case to obtain maximum polymerization activity. This resulted in fairly constant ethylene consumption from run to run, thus eliminating the rate of ethylene dissolution into the liquid phase as a significant variable.

The results show that Cr efficiency declined sharply as the Cr concentration in the catalyst was increased. However, overall catalyst efficiency increased sharply with increased Cr concentration up to 0.75% Cr.

Also shown in Table V is the calculated average distance between Cr ions

TABLE V

Effect of Catalyst Cr Content on Polymerization Activity

Cr in catalyst (wt%)	Distance between Cr ions (Å)	Polyethylene yield (g)	
		Per gram of Cr	Per gram of Catalyst
0.01	160	4,300,000	430
0.03	93	3,000,000	900
0.10	51	1,920,000	1920
0.30	30	1,500,000	4500
0.75	19	1,200,000	8960
1.0	16	936,000	9360
1.5	13	593,000	8900
2.0	11	460,000	9100
4.0	8	207,000	8270
6.0	6.5	98,000	5900

on the surface (300 m^2/g), assuming that there is no clumping of Cr and no formation of dichromate. At 1% Cr, the distance between Cr ions is only 16 Å, and one could expect competition for ethylene on the surface between two polymerization sites spaced this closely. At only 0.01% Cr concentration, the average distance between Cr ions cannot be less than about 160 Å, and competition for ethylene between sites cannot be a factor. At intermediate concentrations, as Cr is added more sites are created, as shown by the

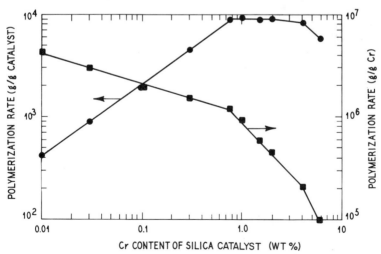

Fig. 7. Effect of Cr content of Chromium–Silica catalyst on Cr efficiency and total catalyst efficiency.

increase in catalyst efficiency. However, the average rate of polymerization at each site decreases. Because the polymer concentration at each site also must decrease, shielding of the site from ethylene by polymer accumulation does not account for the decrease. Apparently, the rate of transfer of ethylene from the liquid phase to the adsorbed phase is the main restraint. The rate of transfer of ethylene from the liquid phase to a catalyst particle is a function of the concentration of ethylene on the surface of the particle. (As stated earlier, the concentration of ethylene in the liquid phase can be considered to be constant in all cases.) When the active site concentration on a particle is increased (by the addition of more Cr), the surface ethylene concentration is lowered and the rate of ethylene transfer to the catalyst particle is thus increased, bringing about decreased site efficiency and increased catalyst efficiency simultaneously, as shown in Fig. 7.

It has been theorized that a very low Cr concentration results in the highest Cr efficiency because there are a limited number of special locations on the silica surface that produce very high activity and these sites are utilized by the Cr ions first. However, this seems unlikely, since the reaction between CrO_3 and silanol groups on the surface appears to be a simple chemical reaction which eliminates water (Section II).

When more than 1% Cr was put on the catalyst (2–6% Cr, Table V), clumping of the Cr occurred and Cr efficiency declined more sharply. The number of active sites apparently leveled out and then decreased, since catalyst efficiency declined. This is believed to be a result of the formation of Cr_2O_3 and possibly dichromate.

These results indicate that the active catalyst site contains a single Cr ion present as silyl chromate before reaction with ethylene. Positive identifications of active sites in heterogeneous catalysts are few [65], but this appears to be a case in point.

Because the polymer produced over this catalyst contains a vinyl group at one end and a methyl group at the other, the question arises as to whether initiation or termination produces the double bond. Unfortunately, no conclusive work along this line has appeared.

Vinyl group formation during termination was suggested earlier by the author [28]. Polymer growth was pictured as follows, the source of the methyl group being shown later:

$$(3)$$

Termination of the growing chain may occur by hydride ion transfer:

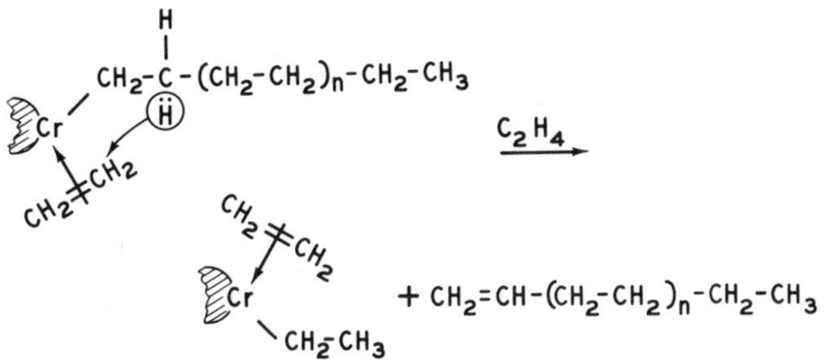

(4)

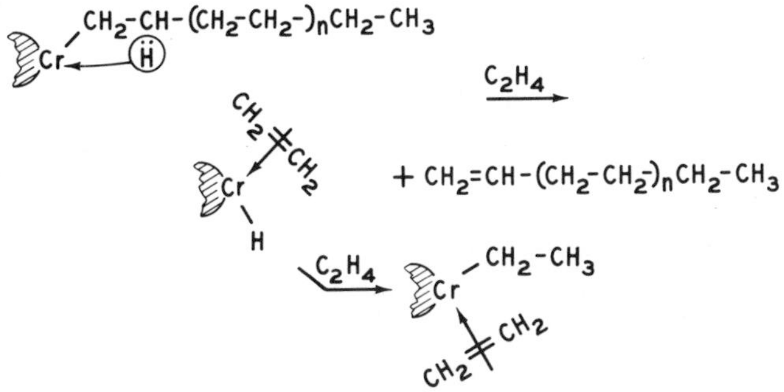

(5)

Transfer of hydride to the active site instead of to the monomer [Eq. (5)] is a possibility. In either case, the methyl group on the starting end and the vinyl group on the terminating end are accounted for.

A slightly different way of viewing the process and also forming a double bond at initiation is proposed as follows:

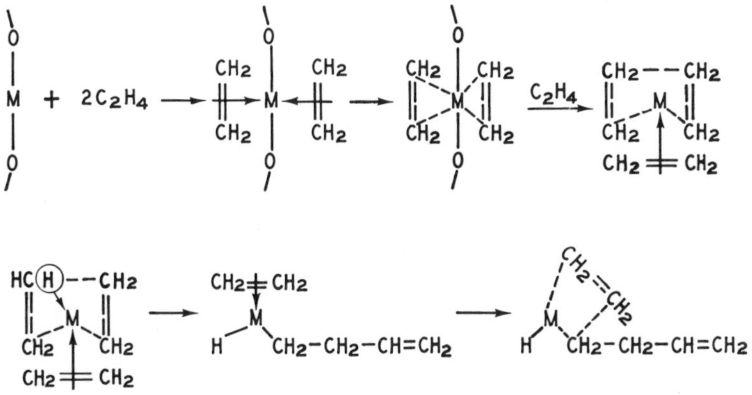

(6)

$$\text{C}_2\text{H}_4 \xrightarrow{\quad} \quad \begin{matrix} \text{CH}_2 \text{\textcircled{=}} \text{CH}_2 \\ \text{M} \\ \text{\textcircled{H}} \rightarrow \text{CH}_2\text{-(CH}_2\text{-)}_n\text{CH=CH}_2 \end{matrix} \xrightarrow{\text{C}_2\text{H}_4} \quad \begin{matrix} & \text{O} & \\ \text{CH}_2 & | & \text{CH}_2 \\ \| \rightarrow \text{M} \leftarrow \| \\ \text{CH}_2 & | & \text{CH}_2 \\ & \text{O} & \end{matrix} \quad + \quad \text{CH}_3\text{-(CH}_2\text{-)}_n\text{CH=CH}_2$$

(The O—M—O bonds are not shown in some steps for simplicity.)

In either case, on contact with ethylene the inevitable oxidation – reduction reaction characteristic of Cr(VI) occurs. A chromium – ethylene complex forms, producing the blue color indicative of interaction between Cr d orbitals and ethylenic π orbitals. In the presence of excess ethylene, ethylene polymerization occurs because of the opening (activation) of double bonds through distortion of the π orbitals.

The addition of α-olefin to the ethylene feed results in the incorporation of α-olefins as follows [from Eqs. (3)–(5)]:

$$\begin{matrix} \text{Cr} \overset{\text{CH}_2\text{-CH}_2\text{-(CH}_2\text{-CH}_2)_n\text{-CH}_2\text{-CH}_3}{\diagup} \\ \text{CH}_2 \text{\textcircled{+}} \overset{\text{CH}}{\underset{\text{CH}_3}{}} \end{matrix} \xrightarrow{\quad} \begin{matrix} \text{Cr} \overset{\text{CH}_2\text{-CH}_2\text{-(CH}_2\text{-CH}_2)_n\text{-CH}_2\text{-CH}_3}{\diagup} \\ \text{CH}_2 \overset{\text{CH}}{=\!=} \underset{\text{CH}_3}{} \end{matrix} \qquad (7)$$

More C_2H_4 units then add. However, termination can leave the α-olefin on the end:

$$\begin{matrix} \text{Cr} \leftarrow \text{\textcircled{H}} \\ \text{CH}_2\text{-C-CH}_2\text{-CH}_2\text{-(CH}_2\text{-CH}_2)_n\text{-CH}_2\text{-CH}_3 \\ | \\ \text{CH}_3 \end{matrix}$$

$$\xrightarrow{\text{C}_2\text{H}_4} \quad \text{Cr} \overset{\diagup\text{H}}{\underset{\text{CH}_2 \text{\textcircled{+}} \text{CH}_2}{}} \quad + \qquad (8)$$

$$\begin{matrix} \text{CH}_2\text{=C-CH}_2\text{-CH}_2\text{-(CH}_2\text{-CH}_2)_n\text{-CH}_2\text{-CH}_3 \\ | \\ \text{CH}_3 \end{matrix}$$

$$CH_2-CH_2-(CH_2-CH_2)_n-CH_2-CH_3$$

$$\underset{\begin{array}{c}Cr\\CH_2\\CH_3\end{array}}{}$$

$$\longrightarrow \quad \longrightarrow \quad \longrightarrow$$

(9)

$$\underset{CH_3}{CH=CH-CH_2-CH_2-(CH_2-CH_2)_n-CH_2-CH_3}$$

The reaction shown in Eqs. (7) and (8) appears to predominate because branched vinyl unsaturation is found in appreciable concentrations in copolymers. However, slight internal unsaturation does appear, especially when the α-olefin content is quite high (above 5 wt%), and can be accounted for by Eq. (9). The transfer of tertiary hydride (compared with the transfer of secondary hydride, when only ethylene is in the chain) can account for the increased termination rate (lowered molecular weight) brought about by the presence of α-olefin in the feed.

The mechanism of long-chain branch formation (at very low concentration) in particle form polymerization referred to in Section IV is pictured as follows. Termination of a normal chain results in the formation of a vinyl group [Eqs. (4) or (5)]. The terminated chain end remains in the vicinity of the active site, since the growing chain crystallizes as it is formed, and the polymer molecule is relatively immobile. Occasionally a terminated chain end (vinyl group) reacts with the active site during the growth of another chain, thus forming a branch approximately equal in length to a typical polymer molecule:

$$\underset{CH_2}{Cr}\overset{CH_2-(CH_2-CH_2-)_nCH_3}{\underset{CH-(CH_2-CH_2-)_{n'}CH_3}{}} \quad \xrightarrow{C_2H_4} \quad \underset{CH_2}{Cr}\overset{CH_2}{\underset{CH_2-CH-(CH_2-CH_2-)_{n'}CH_3}{CH_2-CH_2-(CH_2-CH_2-)_nCH_3}}$$

(10)

Because no evidence of long-chain branching is found in Phillips solution polymers, it supports this hypothesis. That is, in solution polymerization, the terminated chain is in solution and moves away from the active site, preventing a secondary reaction. For the reaction shown in Eq. (10) to occur, the vinyl group must be formed at termination.

Because the molybdenum and vanadium-promoted catalysts used in ethylene polymerization by Amoco Chemicals Corporation [66–68] contained transition metals that are very close in the periodic chart to chromium, similarities in the polymerization mechanisms and in the polymers

TABLE VI

Reducibility of Metal Oxides by CO

Metal oxide on silica–alumina	Temperature at which reduction began
CrO_3, MP 196°C	145–150°C
V_2O_5, MP 690°C	370–375°C
MoO_3, MP 795°C	425–430°C

produced are expected. However, there are important differences. The reported use of hydrogen, sodium, or aluminum alkyls to activate Amoco catalysts reflects the fact that the oxides of Mo and V are more difficult to reduce than those of chromium. The differences in reducibility with carbon monoxide are shown in Table VI. In each case the metal oxide at a 4–5% concentration on silica–alumina was calcined in air at 760°C and then subjected to 1 atm of dry CO at gradually increasing temperatures until reduction, as shown by CO_2 formation and change in color, started.

The data in Table VI show that CrO_3 is reduced at low temperatures by CO, whereas much higher temperatures are required for V_2O_5 and MoO_3. Similarly, ethylene at polymerization temperature does not readily reduce and activate vanadium or molybdenum catalysts, and prereduction or use of a promoter that is a reducing agent is required for high activity. As shown in Table VI, the melting points of the oxides shown increase sharply from Cr to V and Mo. The low melting point of CrO_3 ensures its mobility on the silica surface, ensuring dispersion.

References

1. J. P. Hogan, and R. L. Banks, Belgian Pat. 530,617 (1955); U. S. Patents 2,825,721 (1958); 2,846,425 (1958); 2,951,816 (1960).
2. Alfred Clark, J. P. Hogan, R. L. Banks, and W. C. Lanning, *Ind. Eng. Chem.* **48,** 1152 (1956).
3. H. R. Sailors, and J. P. Hogan, *J. Macromol. Sci. Chem.* **A15** (7) 1377 (1981).
4. G. C. Bailey, and James A. Reid, U. S. Patents 2,381,198 (1945); 2,581,228 (1952); 2,606,940 (1952).
5. M. R. Cines, G. H. Dale, E. W. Mellow, and R. E. Weis, *Chem. Eng. Prog.* **54**(2), 95 (1958).
6. D. E. Berger and R. G. Atkinson, U.S. Pat. 3,078,265 (1863).
7. J. E. Cottle, U.S. Patents 2,952,671 (1960); 3,014,849 (1961); with J. I. Stevens, 3,084,149 (1963).
8. R. F. Dye, U.S. Pat. 2,815,027 (1958); 3,023,203 (1962.)
9. R. R. Goins, U.S. Pat. 2,957,861 (1960); 2,964,512 (1960); 2,936,303 (1960).
10. W. B. Henderson, U.S. Pat. 2,964,516 (1960).
11. E. W. Mellow, U.S. Pat. 3,074,919 (1963).
12. J. J. Moon, and R. M. Hawkins, U.S. Pat. 2,993,599 (1961).
13. E. L. Czenkusch and W. L. Fawcett, U.S. Pat. 3,002,963 (1961).

14. R. G. Wallace, U.S. Pat. 3,056,772 (1962).
15. W. J. Wride, U.S. Pat. 3,055,879 (1962).
16. J. A. Weedman, W. E. Payne, and O. W. Johnson, *Chem. Eng. Prog* **55**(2) 49 (1959).
17. J. W. Davison, and G. E. Hays, *Chem. Eng. Prog.* **54**, 52 (1958).
18. R. V. Jones and P. J. Boeke, *Ind. Eng. Chem.* **48**, 1155 (1956).
19. J. E. Pritchard, R. J. Martinovich, and P. J. Boeke, *Plast. Technol.* **6**, 31 1960).
20. J. E. Pritchard, R. M. McGlamery, and P. J. Boeke, *Mod. Plast.* **37**, 132 (1959).
21. R. V. Jones *Plast. World* **27**(4), 32 (1969).
22. D. C. Smith, *Ind. Eng. Chem.* **48**, 1161 (1956).
23. Alfred Clark and J. P. Hogan, In *"Polythene"* (A. Renfrew and P. Morgan, eds.), 2nd ed., p. 29. Wiley, New York, (1960).
24. G. T. Leatherman, British Pat. 853,414 (1960).
25. *Kirk-Othmer Encycl. Chem. Technol. 2nd ed.* **14**, 253–254 (1967).
26. J. P. Hogan, U.S. Pat. 2,959,578 (1960).
27. Bert Horvath, Unpublished work (1967).
28. J. P. Hogan, *J. Polym. Sci. Part A-1* **8**, 2637 (1970).
29. J. P. Hogan and A. G. Kitchen, U.S. Pat. 3,225,023 (1965).
30. D. R. Witt, U.S. Pat. 3,900,457 (1975).
31. R. E. Dietz, U.S. Pat. 3,887,494 (1975).
32. M. P. McDaniel and M. B. Welch, U.S. Pats. 4,151,122 (1979), 4,177,162 (1979), 4,182,815 (1980).
33. R. E. Campbell, Technical Papers, SPE Reg. Tech. Conf., Houston, February 23–24, pp. 227–241 (1981).
34. J. P. Hogan, D. D. Norwood, and C. A. Ayres, In "Applied Polymer Symposia Series" (H. F. Mark, ed.), Vol. 36. Wiley (Interscience), New York (1981).
35. J. P. Hogan and A. G. Kitchen, U.S. Pat. 3,476,724 (1969).
36. J. P. Hogan, U.S. Pats. 3,130,188 (1964); 3,165,504 (1965).
37. J. P. Hogan, U.S. Pat. 3,362,946 (1968).
38. H. S. Kimble and N. F. McLeod, U.S. Pat. 2,897,184 (1959).
39. F. E. Wiley, U.S. Pat. 2,977,351 (1961).
40. J. P. Hogan and D. R. Witt, *Am. Chem. Soc. Div. Pet. Chem. Prepr.* **24**(2), 377 (1979).
41. D. D. Norwood, U.S. Pat. 3,248,179 (1966); 3,257,362 (1966).
42. Solvay and Cie. Belgian Pat. 570,981 (1958).
43. W. C. Lanning, J. P. Hogan, R. L. Banks, and C. V. Detter, U.S. Pat. 2,970,135 (1961).
44. R. L. Banks, U.S. Pat. 2,860,127 (1958).
45. E. L. Czenkusch and W. L. Fawcett, U.S. Pat. 3,002,963 (1961).
46. K. Wisseroth, *Angew. Makromol. Chem.* **8**(88), 41 (1969).
47. D. M. Rasmussen, *Chem. Eng.* **79**(21), 104 (1972).
48. ASTM Method D-2238-64T (1964).
49. J. C.Randall, *J. Polym. Sci. Polym. Phys. Ed.* **11**, 275 (1973).
50. D. E. Dorman, E. P. Otocka, and F. A. Bovey, *Macromolecules* **5**, 574 (1972).
51. J. C. Randall, *J. Appl. Polym. Sci.* **22**, 585 (1978).
52. M. T. O'Shaughnessy, Unpublished work (1973).
53. J. P. Hogan, C. T. Levett, and R. T. Werkman, *SPE J.* **23**(11), 87 (1967).
54. V. B. Kazanskii, Yu. I. Pecherskaya, and V. V. Voesodskii, *Kinetika i Kataliz* **1**, 257 (1960).
55. D. E. O'Reilly and D. S. MacIver, *Am. Chem. Soc. Div. Pet. Chem. Boston Meeting, April 1959*, p. C-59; *J. Phys. Chem.* **66**, 276 (1962).
56. P. Cossee and L. L. Van Reijen, *2nd Int. Conf. Catal., Paris 1960*, p. 1679.
57. V. B. Kazanskii and Yu. I. Pecherskaya, *Kinet. Catal.* **2**, 417 (1961).
58. L. L. Van Reijen, P. Cossee, and H. J. Van Haren, *J. Chem. Phys.* **38**, 572 (1963).

59. M. A. Waldrop, Unpublished work (1967).
60. H. L. Krauss, *Preprints, 5th Int. Cong. Catal., Miami Beach* Aug. 20–26, 1972, Paper No. 8.
61. R. L. Banks, U.S. Pat. 3,463,827 (1969).
62. R. L. Banks and G. C. Bailey *J. Catal.* **14**(3), 276 (1969).
63. D. W. Walker, and E. L. Czenkusch, U.S. Pats. 3,123,571 (1964); 3,157,712 (1964).
64. B. E. Nasser, Unpublished work (1969).
65. M. Boudart, *Am. Sci.* **57**(1), 97 (1969).
66. E. L. D'Ouville In "Polythene" (A. Renfrew and P. Morgan, eds.), 2nd Ed., p. 35. Wiley (Interscience), New York (1960).
67. O. O. Juveland, E. F. Peters, and J. W. Shepard, *Polym. Prepr.* **10**(1), 263 (1969).
68. E. F. Peters and O. O. Juveland, French Pat. 1,521,017 (1968).

CHAPTER 7

The Evolution of Ziegler – Natta Catalysts for Propylene Polymerization

Specialty Chemical Division
Stauffer Chemical Company
Dobbs Ferry, New York

I. Introduction

The discovery by Karl Ziegler and co-workers in 1953 that certain combinations of transition metal and organometallic compounds converted ethylene to a linear, high-molecular-weight polymer began a flood of research on low-pressure α-olefin polymerization that has not receded today [1, 2]. In 1954, this reaction was extended by Natta [3] to the formation of crystalline polypropylene from propylene, using titanium trichloride and an aluminum alkyl as catalysts. At nearly the same time similar discoveries were made by Vandenberg of Hercules, Baxter of DuPont, Zletz of Standard Oil of Indiana, and Hogan of Phillips Petroleum. In 1963 Ziegler and Natta jointly received the Nobel Prize in chemistry for their work. The commercial significance of this polymerization process is indicated by the billions of pounds of polyolefins manufactured each year and the thousands of patents and articles that have appeared describing numerous variations of the Ziegler–Natta catalyst system. However, most commercial polypropylene producers still employ a titanium halide catalyst (usually $TiCl_3$) in combination with an aluminum trialkyl or diethylaluminum chloride cocatalyst.

Many reviews and articles have attempted to explain the chemistry responsible for the formation of crystalline polypropylene from Ziegler–Natta catalysts. It is intended herein to provide a historical review of the development of commercial Ziegler–Natta catalysts and polymerization processes for polypropylene. For a more comprehensive review of the literature and theories of Ziegler–Natta chemistry, the reader is referred to *Ziegler–Natta Catalysts and Polymerizations* by Boor [4].

II. Polypropylene Markets and Properties

The continuing worldwide research on Ziegler–Natta catalysis is primarily a result of the outstanding growth of the crystalline (or isotactic) polypropylene market. Commercial production of isotactic polypropylene started about 1957. The total polypropylene market grew from less than 50 million lb in 1960 to over 3.9 billion lb in 1981 [5]. One reason for the commercial success of polypropylene is its combination of excellent mechanical and optical properties with a low density. Many grades of homopolymers and copolymers are produced for use in injection molding, film, fiber, filament, blow molding, extrusion, and other applications. A comparison of selected properties of polypropylene and high-density polyethylene is shown in Table I. Polypropylene combines resilience and strength with improved heat

TABLE I

Properties of Polypropylene and High-Density Polyethylene[a]

Property	Polypropylene	High-density polyethylene
Density (g/cm³)	0.902–0.910	0.941–0.965
Melting point (°C)	167	136
Tensile strength (psi)	4300–5500	3100–5500
Flexural elastic modulus (psi × 10⁵)	1.7–2.5	1.0–2.6
Elongation at break (%)	200–700	20–1300
Impact strength, Izod with notch (ft lb/in.)	0.5–2.2	0.5–20.0
Deflection temperature at 66 psi (°F)	200–250	140–190
Hardness, Rockwell	80–110	60–70 (Shore)
Transmittance (%)	55–90	0–40
Haze (%)	1.0–3.5	10–50

[a] Values obtained from *Modern Plastics Encyclopedia*. McGraw-Hill Inc., New York **51**, 558, 560 (1974).

resistance and hardness. These properties make it particularly suitable for the packaging, automotive, and fiber industries. As a result, the total world nameplate capacity for polypropylene grew in 1980 to 15 billion lb with approximately one-third of this capacity in the United States. The major producers in the United States are Amoco, Arco, Exxon, Gulf, Hercules, Northern Petrochemical, USS Chemicals, Phillips, Rexene, Shell, Soltex, and Texas Eastman.

Three stereoisomers of polypropylene, known as isotactic, syndiotactic, and atactic, can be obtained from Ziegler–Natta catalysts. These isomers differ in the relative configuration of the propylene unit resulting from a head-to-tail linkage, as shown in Fig. 1. Actually the isotactic and syndiotactic chains form helixes to accommodate the methyl groups and can crystallize because of their stereoregularity, whereas the more disordered atactic isomer is amorphous. From a commercial standpoint, the properties of the higher melting isotactic isomer (mp 170–175°C) make it more desirable than the syndiotactic isomer (mp 130–135°C). A small market exists in adhesives for the atactic polymer.

Isotactic polypropylene of high molecular weight is insoluble in boiling heptane, whereas the atactic and low-molecular-weight isotactic polymers are soluble [6]. This property provides a convenient method for separating the less desirable forms of polypropylene from the isotactic fraction. Commercial polypropylene generally contains less than 10% polymer soluble in boiling heptane. Depending on the catalyst and cocatalyst used, the compo-

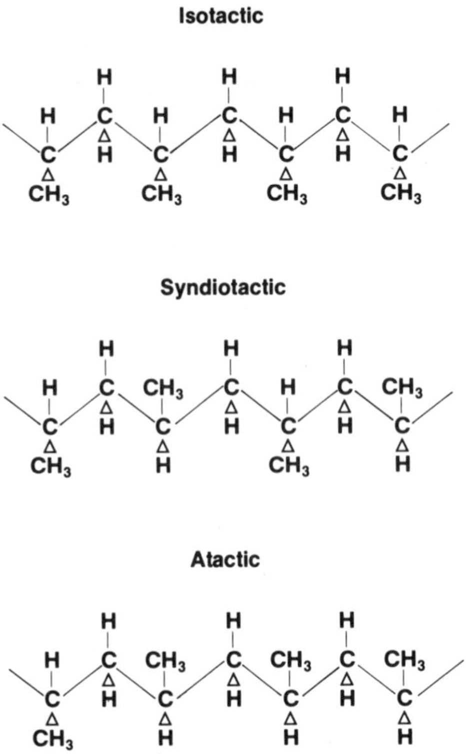

Isotactic

Syndiotactic

Atactic

Fig. 1. Polypropylene structure.

sition of commercial polypropylene is largely high-molecular-weight isotactic with varying quantities of stereoblock (alternating segments of isotactic, atactic, and syndiotactic), atactic, and syndiotactic.

III. Laboratory-Scale Evaluation Procedures

A. REACTOR AND SUPPORT FACILITY FOR POLYMERIZATION STUDIES

The relative quantities of isotactic, atactic, and syndiotactic polypropylene produced by Ziegler–Natta systems are regulated not only by the catalyst and cocatalyst but also by the polymerization conditions. Therefore, before discussing catalyst evolution, a brief summary of polymerization technology is required. First, to acquaint the reader with the terms used in the industry and to outline the basic polymerization procedure, a descrip-

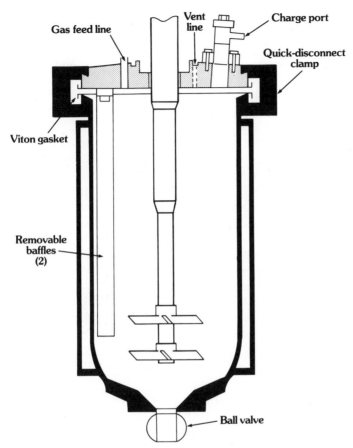

Fig. 2. Laboratory slurry polymerization rector. (From Gold *et al.* [7].)

tion of a slurry-type laboratory test facility will be given. The term "slurry" refers to the fact that the catalyst is suspended or "slurried" in an inert solvent and the propylene is added as a gas. This type of test facility has been used at Stauffer Chemical Company and has the capability for homopolymerization, copolymerization, kinetic measurements, and quick screening of new catalysts [7]. Much of the data reported in this chapter were collected at this test facility.

The facility is divided into a reactor area and a laboratory where polymer processing and catalyst charge preparation occurs. In the reactor area, autoclaves, tested to withstand 500 psig of pressure, are enclosed in separate cells of reinforced concrete block and equipped with metal blast doors, safety glass view ports, and an overhead blast mat. The reactor design is shown in Fig. 2. Reaction temperatures and monomer flows are monitored by thermocouples in the reactor and by both differential pressure cells and mass flow meters in the propylene feed line. Both N_2 and propylene are purified by a supported copper catalyst to remove oxygen and by molecular sieves to remove water. The solvent is purified by passing it successively through columns of activated carbon, silica gel, and molecular sieves. It is then piped directly to the reactors. These absorbents remove unsaturated compounds, sulfur compounds, and water. Solvent purity is routinely monitored by ultraviolet spectroscopy.

In the laboratory area, a glove box is used for preparing the catalyst charge. In testing Ziegler–Natta catalysts, the most important variable affecting results is their extreme sensitivity to moisture and air. Many discrepancies can be found between the results for the same catalyst from laboratories using comparable equipment and polymerization conditions. At Stauffer, we have found that $TiCl_3$-type Ziegler–Natta catalysts maintain their performance for over 1 yr when oxygen and water levels in the glove box are kept below 2 ppm. At levels greater than 2 ppm, catalyst deactivation can occur over time without visible color change in the catalyst. Filtration, evaporation, and drying equipment are used to measure the insoluble and soluble polymer formed. Jacketed Soxhlet units are used to determine the amount of polymer soluble in boiling heptane.

B. LABORATORY POLYMERIZATION PROCEDURE

No standard laboratory test for Ziegler–Natta catalysts has been developed despite more than 25 yr of commercial testing. Such a test would be extremely useful for comparing data published by different laboratories. As a guide for those unfamiliar with testing Ziegler–Natta catalysts, the procedure developed at Stauffer under the reaction conditions listed in Table II will be summarized.

In a typical procedure for polymerization with $TiCl_3$ catalysts, the catalyst charge is placed in a breakable glass ampul in the glove box. The actual amount charged depends on the expected polymer yield. Too high a catalyst loading in the reactor can adversely affect performance at a low polymerization pressure, since monomer transport to the active site may become rate-determining. Alternatively, the chance of catalyst deactivation by poisons is enhanced at a very low catalyst concentration. When new catalysts are evaluated, some experiments for determining the optimal catalyst charge are required. In general the solids content of the resulting polymer slurry should be less than 40%. Beyond this level, heat transfer problems, which can occur during the later stages of polymerization, may cause "hot spots" that may affect the polymerization results.

TABLE II

Laboratory Slurry—Polymerization Conditions[a]

Condition	Value
Reactor size	4.5 liters
Solvent	n-Heptane, 2 liters
Propylene pressure	142 psig
Hydrogen pressure	3.2 psig
Time	4 hr
Temperature	70°C
Catalyst charge, TiCl₃ 1.1	700–800 mg
Et₂AlCl:Ti mole ratio	4
Reactor stirring rate	600 rpm
Typical activity, TiCl₃ 1.1	1150 g PP/g catalyst
Typical isotatic index, TiCl₃ 1.1	90

[a] TiCl₃ 1.1 is activated 3TiCl₃ · AlCl₃ produced by Stauffer. PP, Polypropylene.

To remove poisons adhering to the walls of the reactor, the reactor is treated with a solution of $(C_2H_5)_3Al$ before charging. After this alkyl solution is discharged, the solvent, cocatalyst, and catalyst are added in that order. Some of the aluminum alkyl cocatalyst serves as a "sweetener" to remove traces of impurities from the solvent. The fraction of alkyl consumed by impurities varies with the reactor system and solvent. In the Stauffer laboratory reactors, a solvent with 100 ppm water requires twice the normal amount of aluminum alkyl cocatalyst to achieve the activity found when only 2 ppm water is present in the solvent.

After the catalyst components have been charged and agitation is started to break the ampul, the reactor temperature is raised to ∼15 degrees below the desired polymerization temperature. At this point hydrogen (99.99 purity) is added for molecular weight control, and propylene is added on a demand basis at constant pressure. After an initial exotherm, the reactor is maintained at the polymerization temperature. At the end of the polymerization interval, the propylene flow is stopped and the reactor is cooled and vented.

C.   DETERMINING CATALYST ACTIVITY

After the reactor has been vented, the contents are quenched with methanol and discharged into a container of isopropanol. This alcohol treatment deactivates the catalyst and removes inorganic residues in a process known as deashing. The resulting polymer is separated from the solvents by filtration, washed with more alcohol, dried, and weighed. The polymer soluble in the solvent is determined by evaporation of an aliquot of that solution. Catalyst activity is then determined by

$$\text{catalyst activity} = \frac{\text{wt. of insoluble polymer} + \text{wt. of soluble polymer}}{\text{wt. of catalyst}}. \tag{1}$$

D. MEASURING THE ISOTACTIC INDEX AND XYLENE SOLUBLES OF THE POLYMER

The amount of isotactic polymer is represented by a number called the isotactic index. To determine this value, an aliquot of the dried polymer is further extracted with boiling heptane in jacketed Soxhlet extractors for 3 hr. The amount remaining in the extraction thimble is considered to be isotactic ["% isotactic" in Eq. (2)]. The isotactic index is then defined according to

$$\text{isotactic index} = \frac{(\% \text{ isotactic}) (\text{wt. of insoluble polymer})}{\text{wt. of insoluble polymer} + \text{wt. of soluble polymer}}. \tag{2}$$

Other tests besides extraction in boiling heptane are commonly used by commercial polypropylene producers to gauge the amount of isotactic polymer formed. These tests range from instrumental analyses such as infrared spectroscopy and ^{13}C NMR to the determination of polymer solubility in solvents other than heptane. One popular test involves complete dissolution of the polymer in boiling xylene and recrystallization under controlled conditions. The atactic polymer should remain in solution and is called the xylene solubles. Theoretically, any atactic chains entangled in the crystalline isotactic are more free to dissolve, whereas this may not be the case in a simple heptane extraction. Generally, higher xylene solubles correspond to a lower isotactic index. Care must be exercised, however, to perform the dissolution in xylene under inert atmospheric conditions. At high temperatures, polypropylene can oxidize and the xylene solubles results can be affected.

E. CONTROLLING POLYMER MOLECULAR WEIGHT

The polymer produced under the conditions listed in Table II, using the above operating procedure, has an average molecular weight M_w ranging from 300,000 to 500,000. Without the addition of hydrogen, the molecular weight can approach 10^6. It is postulated that hydrogen competes with the propylene for the active site to form a Ti—H bond and thereby terminates the polymerization process [8].

However, most evidence appears to support the view that the active site is not destroyed by this reaction, and hydrogen actually serves as a transfer agent. The molecular weight distribution of polypropylene chains produced from Ziegler–Natta catalysts is broad. This broadening may result either from different polymerization rates associated with a distribution of types of active sites or from diffusional limitations imposed on the propylene by encapsulation of the catalyst by the polymer.

The molecular weight and molecular weight distribution can be measured by gel permeation chromatography by comparison with standards. However, because the polymer molecular weight affects the physical properties of the polymer, various other more rapid techniques have been applied to approximate the molecular weight value. One of the more popular techniques involves determination of the viscosity in tetralin at several polymer concentrations. The average molecular weight M_w can then be related to the intrinsic viscosity. One relationship between viscosity and M_w that can be applied for M_w up to 650,000 is [9]

$$\text{intrinsic viscosity} = 0.80 \times 10^{-4} M_w^{0.80}. \tag{3}$$

IV. Effects of Polymerization Variables

A. COCATALYSTS

Besides the form of the catalyst itself, the most important variable in the polymerization of propylene is the cocatalyst. Table III shows the effect of metal alkyl on catalyst activity and heptane solubles in a low-pressure slurry polymerization. These results appear to support the findings of others that

(1) The stereospecificity decreases with increasing size of the alkyl group of R_3Al.

(2) The stereospecificity increases with the size of the halogen when present in R_2AlX.

A third effect of the metal alkyl was noted [10] when it was observed that the amount of polymer insoluble in boiling heptane decreased as the ionic radius of the metal increased. This is shown in Table IV. These alkyl effects appear to support the concept that the metal alkyl cocatalyst is actively involved in the polymerization mechanism. One possible mechanism proposed involved insertion of the olefin into the Al—C bond present in a bimetallic complex presumed to form from the titanium halide and aluminum alkyl [11].

This bimetallic mechanism was supported by the discovery through

TABLE III

Effect of the Cocatalyst on Activity and Isotactic Index[a]

Cocatalyst	Activity (g PP/g catalyst)	Heptane-insoluble polymer (%)
Et$_2$AlF	251	97.7
Et$_2$AlCl	159	96.4
Et$_2$AlBr	105	96.7
Et$_2$AlI	42	98.1
Et$_3$Al	346	70.5
(n-Pr)$_3$Al	~400[b]	68.6
(n-Bu)$_3$Al	~400[b]	57.7
(n-Hexyl)$_3$Al	~400[b]	54.0
(n-Octyl)$_3$Al	~400[b]	50.3
(n-Decyl)$_3$Al	~400[b]	41.1

[a] Polymerization conditions: 35 psig C$_3$, no H$_2$, 55°C, 3.5 hr, Al:Ti mole ratio 2.8, n-heptane solvent, and TiCl$_3$ 1.1 as the catalyst. PP, polypropylene.
[b] Polymerization terminated early because of polymer buildup.

radiolabeling experiments that the alkyl group of the aluminum cocatalyst became attached to the polymer chain [12]. However, as we shall see in Section VI.C, this finding can also be explained by a monometallic mechanism involving a Ti—C bond as the propagation center.

B. SOLVENT PURITY

The polymerization solvent is important in slurry reactions because of the different solubility of propylene in various solvents and the varying solubility of the atactic polymer. However, of equal importance is the level of contaminants in the solvent and the propylene. Substances known to poison

TABLE IV

Isotacticity as a Function of Metal Alkyl[a]

Metal alkyl	Ionic radius of metal (Å)	Polypropylene insoluble in boiling heptane
BeEt$_2$	0.35	94–97
AlEt$_3$	0.51	80–90
MgEt$_2$	0.66	78–85
ZnEt$_2$	0.74	30–40

[a] Propylene polymerization at 75°C, 2.4 atm, α-TiCl$_3$ as the catalyst. Natta [10]

the catalyst include water, oxygen, carbon monoxide, carbon dioxide, allene, acetylene, carbon oxysulfide, and organic sulfur compounds. For the best catalytic performance, it is important to maintain the concentration of most of these poisons as low as possible—usually less than a few parts per million. However, the action of each poison is not always predictable. As an example, the effect of the water level in heptane on a commercial TiCl₃ catalyst is shown in Table V. In this case, although activity decreases directly with increasing water concentration, a small improvement in the isotactic index can be realized when 50 ppm of water is present. Commercial polypropylene plants usually employ two techniques for the purification of solvent. Much of the water is removed by contact with molecular sieves. However, traces of water remaining in the solvent after the molecular sieves treatment can be eliminated by adding small amounts of aluminum alkyl. Also, propylene is generally passed over molecular sieves to eliminate water.

C. HYDROGEN PRESSURE

For TiCl₃ systems in a slurry reactor, the polymerization rate of the catalyst generally increases with higher temperatures and propylene pressures, but the stereospecificity decreases. When H_2 is present, the effect on catalyst performance is not so straightforward. This is illustrated in Table VI for a catalyst prepared in the laboratory from $3TiCl_3 \cdot AlCl_3$ and an organic promoter. In this series of polymerization trials, the hydrogen was added in the beginning and not continuously as in commercial reactors. The isotactic index experiences a small initial drop upon the introduction of hydrogen but remains relatively constant with increasing hydrogen concentration until very high levels of hydrogen are reached. The isotactic index then begins to decrease steadily. In contrast the activity increases directly with hydrogen pressure until very high levels of hydrogen are present, and then it

TABLE V

Effect of Water in the Polymerization Solvent[a]

Water (ppm)	Activity (g PP/g catalyst)	Isotactic index
2	1029	89.7
27	975	89.1
52	762	91.8
102	594	90.6

[a] Polymerization conditions: 142 psig C₃, 3.2 psig H₂, 70°C, 3 hr, Diethylaluminum chloride: Ti mole ratio 4, n-heptane solvent, TiCl₃ 1.1 as the catalyst. PP, Polypropylene.

TABLE VI

Catalyst Performance in the Presence of Hydrogen[a]

H$_2$ (psig)	Activity (g PP/g catalyst)	Isotactic index
0	996	96.7
3	1122	95.8
9	1178	95.7
15	1318	91.6
20	1124	90.6
25	972	85.1

[a] Polymerization conditions: 142 psig C$_3$, 70°C, 3 hr, DEAC:Ti ratio 3.9, n-heptane solvent, laboratory milled 3TiCl$_3$ · AlCl$_3$ plus camphor as the catalyst. PP, polypropylene.

too begins to decrease. The competition between propylene and hydrogen for the active site may explain this falloff in activity at high levels of hydrogen.

V.   Evolution of Polymerization Processes

A.   SLURRY POLYMERIZATION

The most widely practiced commercial polymerization process is the slurry technique. In a typical commercial process, the solid catalyst and liquid cocatalyst are slurried in a hydrocarbon solvent. Solvents such as isobutane, pentane, hexane, and xylene are commonly used. This catalyst slurry may be stirred in holding tanks from hours up to days, so good catalyst stability and attrition resistance are desirable. Other components may be added to enhance the stereospecificity of the catalyst. The catalyst slurry, solvent, hydrogen, and propylene are fed to the reactor, and polypropylene is removed continuously. Excess monomer is flashed from the polymer slurry and recycled. Deashing is accomplished by treating the polymer with reagents such as alcohol, HCl, water, and NaOH. If necessary, the polymer is further treated to remove atactic polypropylene. After separation from the solvent and drying, the powder obtained from this process is usually extruded into pellets. Various antioxidant and ultraviolet light stabilizers are added before the extrusion step.

Because this is a multiphase system, stir rates of ∼ 500 rpm and higher are commonly employed during polymerization to ensure good catalyst suspension and mixing of components. Although catalyst productivity is reduced because of the low pressures (usually 50–150 psig) and low temperatures

(50–75°C) employed, the isotactic index is generally above 90. However, for many slurry processes, extraction to remove atactic is still required to obtain an isotactic index of 96–97, the isotactic index suitable for most commercial applications. Thus for the slurry process considerable capital and operating expense are involved in solvent purification and recycle, wastewater treatment, polymer extraction, and deashing. Some of the major producers in the United States who presently use the slurry process are Hercules, Amoco, Exxon, and USS Chemicals.

B. SOLUTION POLYMERIZATION

The main feature of the solution process is the use of polymerization temperatures ranging from 110 to 150°C. This is done to keep the polymer soluble in the solvent. Catalyst residues can thereby be removed by filtration of the hot polymer solution. In this way the expenses involved in deashing can be avoided, and a very pure polypropylene is obtained. The polymer itself is obtained by crystallization from the solution and centrifuging. Usually pressures near 1000 psig are combined with the high polymerization temperature to reduce the reactor residence time and to increase throughput. In Ziegler–Natta systems, however, greater amounts of the normally undesirable atactic polypropylene are produced when the polymerization temperature is increased. Therefore, unless a favorable market position exists in atactic applications, the solution process is less competitive economically compared to the other processes. The only major producer of polypropylene in the United States using solution polymerization is Texas Eastman.

C. BULK POLYMERIZATION

The bulk or liquid monomer processes use liquid propylene as both the solvent and the reactant. This is accomplished by operating at a low temperature (55–80°C) and moderate pressure (300–400 psig) to keep the propylene below its critical point. Both Rexene and Phillips have developed reactors that utilize this concept.

In the Rexene process, all components are fed to a stirred batch reactor while polypropylene is removed continuously. Much of the heat of reaction is removed by boiling liquid propylene in a downflow condenser. Excess propylene is flashed from the polymer, condensed, and recycled back to the reactor. However, this recycle causes a potential problem from the buildup of oligomerization products and contaminants in the propylene if a subsequent purification step is not included. The polymer is then treated with alcohols and extracted with solvents as in slurry polymerization to remove catalyst residues and atactic polypropylene, depending on the efficiency and stereospecificity of the catalyst system.

The Phillips bulk reactor is in the shape of a loop in which the reactants are circulated at high speed. Again polypropylene is continuously removed and subsequently treated to remove catalyst residues and atactic polymer. This loop process reportedly combines improved heat transfer through convection without periodic shutdown that can be required because of polymer deposition on the reactor walls [13]. However, both the Phillips and Rexene systems possess the advantages of increased throughput because of their high polymerization rate compared to those of conventional slurry processes. There is also a reduced investment for solvents and purification systems. Also, in many cases the isotactic content of the polymer is improved over that found for slurry polymerization.

D. GAS PHASE POLYMERIZATION

The latest development in propylene polymerization reactor design is a fluidized bed in which the catalyst is suspended in a stream of propylene gas. Fluidization can be assisted by mechanical means. Although no solvent is used, the catalyst is usually injected as a hydrocarbon slurry and frequently is supported on an inert support such as polypropylene. Again, this type of polymerization has economic advantages, since no polymerization solvents are involved and the polymer is obtained continuously without a need for centrifuges or other separation equipment. No fractionation is required for the propylene recycle loop. Heat removal is assisted by vaporization of the propylene, which is injected as a liquid, but temperature control and fusion of particles are constant problems.

The original developments in this field were made by Badische Anilin- & Soda-Fabrik (BASF), and the early problems with low isotactic content have been solved. Polymerization pressures and temperatures are similar to those found in bulk polymerization. Although considerably less equipment is required to support the fluid bed reactor, because solvent recycle and purification have been eliminated, the reactor is difficult to scale up as a result of the particle fusion and temperature control problems. Northern Petrochemical and Amoco are the only major producers in the United States currently using gas phase polymerization. The Amoco process is somewhat different from the BASF-type reactor in that a horizontal bed is employed.

VI.   Commercial Catalyst Development

A.   GOALS

As we have seen, the polymerization process and variables greatly influence the performance of Ziegler–Natta catalysts. However, the catalyst itself plays the most important role in determining the activity and isotactic index. As commercial application of these catalysts became a reality, two major goals were established to improve the economics of the process. These goals were to provide a catalyst with (1) activity high enough to avoid deashing and (2) an isotactic index sufficient to eliminate the extraction of atactic polymer. For most commercial applications of polypropylene, the first goal would be met if the Ti and Cl residuals in the polymer were below 5 and 20 ppm, respectively. To meet the second goal an isotactic index of 96–97 would be desirable. In pursuit of these goals, the development of commercial Ziegler–Natta catalysts has spanned 25 yr at five plateaus of performance. This is shown in Table VII. A description of each of these periods follows.

B.   THE *in Situ* CATALYST

The earliest commercial Ziegler–Natta catalyst system involved preparation of $TiCl_3$ directly in the polymerization reactor from $TiCl_4$ and aluminum alkyl compounds. The exact form of $TiCl_3$ resulting from the reduction of $TiCl_4$ depends strongly on the reaction conditions, solvents, and the ratio of reductant to $TiCl_4$. Examples of various preparative routes to $TiCl_3$ are

TABLE VII

Polypropylene Catalyst Development Periods

Development stage	Earliest year of commercial use	Relative activity (wt/wt Ti)	Isotactic content (%)
In situ production catalyst	1957	1	75–85
External production catalyst	1960	10	85–89
Promoted catalyst	1970	12	93–98
High-surface-area catalyst	1976	40	94–96
Supported catalyst	1977	1000	89–94

given in Table VIII. Under the conditions employed during reduction in the polymerization reactor, only the brown beta phase of TiCl₃ would be formed. This phase of TiCl₃ has a fiber-type structure and is composed of linear chains of titanium atoms linked by chloride bridges. This material had moderate activity and low stereospecificity and thus was not very attractive economically because extensive extraction of the polymer is required to remove atactic polypropylene. However, it was quickly discovered that the more stereospecific purple TiCl₃ was formed upon thermal treatment of the beta phase. Although this procedure was not well suited to a continuous polymerization process, the effect of the crystalline form of TiCl₃ on the activity and the isotactic index was underscored. A comparison of the performance of the four phases of TiCl₃ is shown in Table IX. Clearly the beta form is inferior to the purple forms of TiCl₃ in the production of isotactic polymer.

C. EXTERNALLY PRODUCED CATALYSTS

In the early 1950s Stauffer Chemical Company produced titanium trichloride from the reduction of TiCl₄ with aluminum or hydrogen. When aluminum was used as the reductant, AlCl₃ was introduced into the TiCl₃ lattice in an isomorphous manner, as described by Natta et al. [17]. The reaction proceeds according to

$$3TiCl_4 + Al \xrightarrow{\Delta} 3TiCl_3 \cdot AlCl_3 \tag{4}$$

In the case of aluminum it is beneficial to include a small amount of AlCl₃ to initiate the reaction. It was also found that the physical state of the alumi-

TABLE VIII

Preparation of Titanium Trichloride

Reaction	Reference
$TiCl_4 + H_2 \xrightarrow{800°C} \alpha\text{-}TiCl_3$	Klemm and Krose [14]
$TiCl_4 + \text{organoaluminum compound} \xrightarrow{\text{low temp.}} \beta\text{-}TiCl_3 \cdot xAlCl_3$	Natta et al., [15]
$TiCl_4 + Al \longrightarrow \gamma\text{-}TiCl_3 \cdot xAlCl_3$	Tornquist and Langer [16]
$\alpha\text{-}TiCl_3 + AlCl_3 \xrightarrow{\text{grinding}} \delta\text{-}TiCl_3 \cdot xAlCl_3$	Natta et al., [17] Tornquist and Langer [16]

TABLE IX

Crystalline Phases of Titanium Trichloride[a]

Phase	Color	Crystal form	Activity (g PP/g catalyst)	Isotactic index
Alpha	Purple	Hexagonal close pack	56	82.4
Beta[b]	Brown	Linear chains	561	76.2
Gamma[c]	Purple	Cubic close pack	102	86.1
Delta[c]	Purple	Hexagonal plus cubic close pack	1150	89.8

[a] Polymerizations conditions: 142 psig C_3, 3.2 psig H_2, 70°C, 4 hr, DEAC:Ti mole ratio 4, n-heptane solvent. PP, Polypropylene.
[b] Prepared from $TiCl_4 + (C_2H_5)_2AlCl$ at 0°C.
[c] Contains $AlCl_3$.

num metal was important. Following the finding that this type of $TiCl_3$ was an effective catalyst for the polymerization of propylene, Stauffer was able to offer commercial quantities to customers in the United States. These sales marked the beginning of the external production catalyst period.

When either aluminum or hydrogen is used as the reductant, the purple titanium trichloride produced exists in the gamma or alpha phase, respectively (see Table VIII). These phases have a layered structure with the titanium atoms in an octahedral environment. The two phases differ in the relative crystalline packing of the two-dimensional layers. The alpha form is hexagonal close-packed, and the gamma form is cubic close-packed. Although both these phases are active in producing a relatively high percentage of isotactic polypropylene, their productivity is low. However, it was discovered that these phases could be activated by intense grinding in ball mills or similar devices. This grinding action produced a fourth phase of titanium trichloride known as the delta phase. This form is characterized by a randomly alternating sequence of layers, one corresponding to the alpha form and the other to the gamma form.

For most commercial catalysts that are activated by grinding, the surface area increases from less than 5 m^2/g before grinding to greater than 20 m^2/g. Dry grinding of 3γ-TiCl$_3$ · $AlCl_3$ increases the activity, as demonstrated in Table X. Because the activity continues to increase despite a drop in surface area on prolonged grinding, catalyst efficiency is not a function of surface area alone. Examination of the x-ray diffraction data for these catalysts shows that the line at 2.70 Å broadens, becomes diffuse, and shifts to 2.67 Å after 96 hr of grinding. This suggests formation of the delta phase during the extended grinding.

A range of catalyst particle sizes is obtained as a result of the grinding

TABLE X

Effect of Laboratory Grinding of 3TiCl$_3 \cdot$ AlCl$_3$[a]

Grinding time (hr)	Surface area (m^2/g)	Activity (g PP/g catalyst)
0	<5	35
5.5	5	62
27.5	—	84
49.5	45	100
69.5	—	120
84.5	—	142
96	34	148

[a] Polymerization conditions: 35 psig C$_3$, no H$_2$, 55°C, 3.5 hr, DEAC:Ti mole ratio 2, *n*-heptane solvent. PP, Polypropylene.

process. Usually the diameter of the particles extends from the submicrometer region to over 200 μm. One convenient method for determining the actual particle size distribution is to conduct a polymerization and to separate the polymer particles by sieving into a polymer particle size distribution. The catalyst particle size distribution can then be found because the polymer replicates the catalyst according to Eq. (5).

$$P_s = C_s[(D_1/D_2)(CA)(C_{7i})]^{1/3}, \qquad (5)$$

where P_s is the polymer diameter in micrometers, C_s the catalyst diameter in micrometers, D_1 the density of the catalyst, D_2 the density of the polymer, CA the activity in grams of polypropylene per gram of catalyst, and C_{7i} the percentage of heptane-insoluble polymer.

For the efficient operation of polymer processing equipment, the production of fine and coarse polymer fractions should be minimized. Also, the presence of polymer fines in polymer storage bins can lead to a safety hazard because of the possibility of an explosion. Thus, one stage in the external catalyst manufacturing process is restriction of the corresponding catalyst fine (diameter <10 μm) and coarse (diameter >150 μm) fractions. The particle size distribution of a typical commercial catalyst is shown in Fig. 3.

Some of the physical transformations that occur upon grinding have been studied at Stauffer by the examination of scanning electron micrographs. Unground 3TiCl$_3 \cdot$ AlCl$_3$ consists of aggregates of primary hexagonal crystallites that are remarkably uniform in size. These crystallites are 0.1–0.5 μm in length and 0.05-μm thick. On grinding, they are transformed into dense, irregularly shaped globules that are lamellar and have terraced, lacy edges. This morphology suggests that the bonds between lamellae are weaker than the bonds within individual lamella. The lamellae are from 200

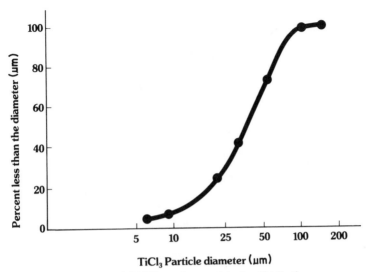

Fig. 3. Commercial TiCl₃ catalyst particle size distribution.

to 500 Å thick, and the globular particles range in size from 0.1 to 170 μm in diameter. Some small particles, which are 2–10 μm in diameter, adhere to larger particles and in some cases have a hexagonal shape. However, these hexagons are not likely to be the original crystallites because they are at least four times the size of the primary crystallites. On the larger particles there are cracks that eventually may lead to breaking of the particles during polymerization.

During the grinding process, defect sites are presumably formed as a result of sliding at the metal interfaces of the Cl—M—Cl double layers. In this case, surface titanium atoms are present in an octahedral environment, containing coordination vacancies. This physical arrangement appears to agree with the active site postulated by Cossee in 1964 [18] in his proposed mechanism for the polymerization process for ethylene, as shown in Fig. 4. In this reaction sequence one chlorine attached to titanium is replaced by an alkyl group from the cocatalyst to form a Ti—C bond. The propylene molecule, which is π bonded to the titanium at a vacant coordination site, then inserts into the Ti—C bond. Stereoregular polymerization is accomplished by exchange of the newly regenerated vacant site with the polymer chain. According to Cossee only transition metals with one to three unpaired d electrons and $3d$ orbitals with energy levels falling between the bonding and antibonding levels of the olefin would be active.

Although many mechanisms describing the polymerization of propylene involving monometallic, bimetallic, bound radical, and bound anionic

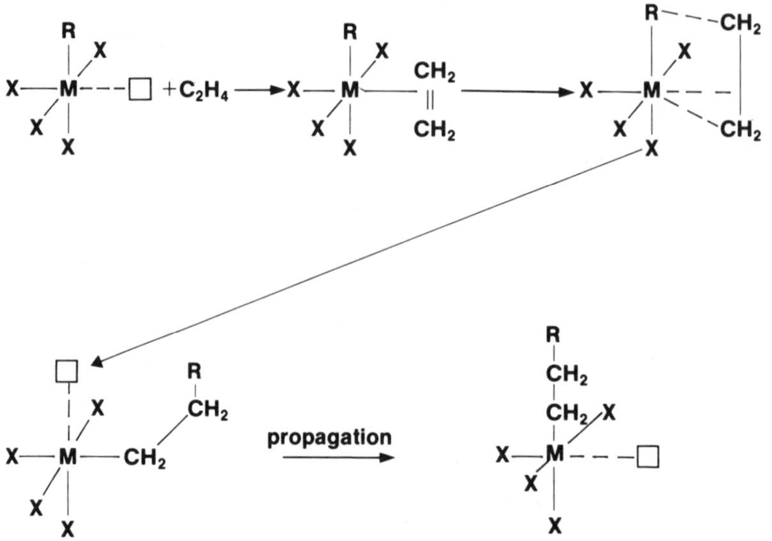

Fig. 4. Cossee's monometallic mechanism. (From Cossee [18].)

active sites have been proposed, no definitive evidence of one mechanism is available. However, in choosing between propagation at the Al—C bond or the Ti—C bond, most experimental evidence supports the latter. In the production of isotactic polypropylene it appears that the steric restrictions imposed by a heterogeneous system are most effective. No homogeneous system has yet been developed that produces a high percentage of isotactic polypropylene in good yield.

TABLE XI

Specifications for TiCl₃AA Type 1.1ᵃ

Variable	Specification
TiCl$_3$ (wt%)	76.0–78.0
TiCl$_4$ (wt%)	0.8 maximum
Total iron (wt%)	0.05 maximum
Activity (g PP/g catalyst)	1000 minimum[a]
Isotactic index	88.0 minimum[a]
Typical particle size (wt% less than)	
5 μm	10 maximum
10 μm	15 maximum
30 μm	44 maximum
100 μm	94 maximum
150 μm	100 maximum

[a] Polymerization conditions: 142 psig C$_3$, 3.2 psig H$_2$, 70°C, 4 hr, DEAC:Ti mole ratio 4, n-heptane solvent. PP, Polypropylene.

With the introduction of commercial external production catalysts, the responsibility for quality control of the catalyst was transferred from the polypropylene producer to the catalyst manufacturer. Several impurities must be minimized to ensure the high performance of these extremely air-sensitive compounds. The first consideration during manufacture and shipment must be the exclusion of air and moisture to avoid the formation of contaminants such as TiO_2 and $TiOCl_2$. Because iron tends to discolor the polymer, provision must be made for the removal of traces of iron, which can result from the extensive grinding used to activate the catalyst. The $TiCl_4$ level must be as low as possible because the interaction of residual $TiCl_4$ with the cocatalyst in the polymerization reactor produces small amounts of the less stereospecific β-$TiCl_3$. The specifications for a typical commercial external production catalyst are given in Table XI.

D. PROMOTED EXTERNALLY PRODUCED CATALYSTS

Externally produced catalysts provided activity and an isotactic index sufficient for commercial application on a continuous-polymerization basis and indeed are still used today in slurry, bulk, and gas phase processes. However, in most cases expensive deashing and atactic removal steps are still required to produce acceptable polypropylene. In fact, atactic formation is doubly disadvantageous because of both removal costs and the loss of monomer that could be converted to the more profitable isotactic polymer on recycling. Thus, for companies without markets for atactic polymer, maximizing isotactic formation is highly desirable even if some activity is sacrificed. Early in the external production period, the addition of electron donors (or Lewis bases) to the catalyst slurry was found to improve the stereospecificity. However, in many cases the activity was adversely effected. Subsequently, electron donors were incorporated into the $TiCl_3$ by the catalyst manufacturer in amounts ranging from 1 to 20% of the catalyst weight. For some systems, this method was superior to addition of the electron donor to the reaction mixture before polymerization. A comparison of the two methods is illustrated for hexamethylphosphoric triamide in Table XII. This use of electron donors in combination with $TiCl_3$ began the promoted catalyst era.

Although thousands of electron donors have been claimed, few are commercially useful. Besides improved catalytic performance, some criteria a successful electron donor must meet are moderate cost, approval as an indirect food additive for food wrap applications, minimal disruption of polymer properties, and compatability with polymer deashing reagents and stabilizer additives. One class of electron donors developed at Stauffer

TABLE XII

Effect of Electron Donor on 3TiCl$_3$ · AlCl$_3$[a]

Catalyst	Activity (g PP/g catalyst)	Isotactic index
3TiCl$_3$ · AlCl$_3$ (A)	531	90.1
A + 0.1 mol HMPT[b]	531	92.5
A + 0.3 mol HMPT[b]	499	93.0
A + 0.1 mol HMPT, ground	484	95.4

[a] Polymerization conditions: 87 psig C$_3$, 1.5 psig H$_2$ 70°C, 2 hr; DEAC:Ti mole ratio 3.3, n-heptane solvent. PP, Polypropylene.
[b] HMPT, Hexamethylphosphoric triamide added as a n-heptane solution to the polymerization reactor.

Chemical Company that meets these requirements is the monoterpenic ketones. The performance of promoted catalysts with terpenic ketones as electron donors is presented in Table XIII. From these data, it is clear that both the activity and isotactic index can be improved by proper choice of electron donor.

The exact role of the electron donor in increasing the activity and/or isotactic index is not clearly understood and probably varies with the catalyst system used. However, because of the heterogeneous nature of the TiCl$_3$ system, a distribution of active sites with varying stereoregulating and polymerization rate properties probably is present. It is plausible that electron donors interact in both electronic and steric capacities to deactivate nonstereospecific sites, convert nonstereospecific sites to stereoregulating sites, and assist in the generation of new active sites during polymerization.

Electron donors used in situ in the polymerization reactor and ground with TiCl$_3$ have been effective in improving the conversion of propylene to isotactic polymer and are used extensively in the industry. In some instances

TABLE XIII

Monoterpenic Ketones in Promoted Catalysts

Electron donor	Activity (g PP/g catalyst)[a]	Isotactic index
None	796	89.6
d,l-Fenchone, 3.5%	702	92.7
Thujone, 5.1%	1014	92.3
Menthone, 3.1%	1115	92.1
Camphor, 5.1%	999	93.3

[a] Polymerization conditions: 142 psig C$_3$, 3.2 psig H$_2$, 70°C, 3 hr; DEAC:Ti mole ratio 3.3, n-heptane solvent. PP, Polypropylene. (From Arzoumanidis [19].)

the isotactic polymer content level needed to eliminate atactic removal can be achieved. Unfortunately, in most cases, the titanium level in the catalyst exceeds 20%. With an activity of 1000–2000 g of polypropylene per gram of catalyst, without deashing over 120 ppm of titanium remains in the polymer. Because for many commercial applications the Ti level in polypropylene should not exceed 5 ppm, deashing is still required for electron donor-promoted catalysts. Yet, for many commercial plants the promoted externally produced catalyst is economically superior to the unpromoted externally produced catalyst because of the reduction in atactic removal expenses and more efficient use of propylene monomer.

E. THE HIGH-SURFACE-AREA CATALYST

The fourth stage in Ziegler–Natta catalyst development addressed the deashing problem by using high-surface-area catalysts to increase the active site population and to improve activity. During this period, Solvay developed catalysts with surface areas exceeding 75 m²/g by careful control of reaction conditions in the sequence of reactions shown in Fig. 5. In this procedure one of the roles of the complexing agent, such as dibutyl ether, is the removal of aluminum compounds from the brown $TiCl_3$ initially produced. The more stereospecific purple $TiCl_3$ is obtained after further treatment with $TiCl_4$. In separate developments, this multistep process for obtaining a high-surface-area purple $TiCl_3$ was replaced by a single-step process. In a one-step process developed by Shell International, solutions of both $TiCl_4$ and aluminum alkyl are complexed with very specific ratios of an ether before mixing [21]. A purple $TiCl_3$ precipitate is then produced when these two solutions are mixed under the proper conditions.

Compared to externally produced catalysts and promoted systems, a 200–400% improvement in activity with no loss in isotactic index can be achieved with high-surface-area catalysts. Polymerization results for typical systems are given in Table XIV. Besides increased polymerization performance, these catalysts offer other advantages. A characteristic of these precipitated catalysts is the production of spherical catalyst particles with a narrow size distribution. A typical catalyst may have over 90% of its particles between 25 and 35 μm in diameter. Because the polymer particles replicate the catalyst particles, a narrow polymer particle size distribution is also found. Uniform particle size with essentially no fine or coarse fraction is desirable for improved handling in polymer processing equipment downstream from the reactor. Theoretically, the expensive steps of extrusion and pellet formation could be avoided if polymer spheres of the correct size were produced. However, because the polymer stabilizer package is generally

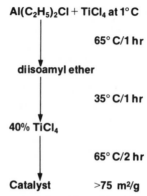

Fig. 5. High-surface-area catalyst. (From Hermans and Henrioulle [20].)

added to the powder just before the extruder, an effective method for adding these components must be devised. Another disadvantage of these systems was discovered during their early development when storage stability problems were encountered. Despite an apparent solution to this problem, the market for high-surface-area catalysts is mainly captive, wherein the catalyst manufacturing equipment is on site with the polymerization facilities.

Even with a polymerization activity of 4000 or 5000 g of polypropylene per gram of catalyst, high-surface-area catalysts still require deashing facilities because, again, the titanium level in the catalyst remains above 20%. Thus the titanium level in the polymer greatly exceeds the 5-ppm target.

F. SUPPORTED TITANIUM CATALYSTS

A solution to the deashing problem was proposed in the late 1960s with developments in supported titanium catalysts. This fifth stage of catalyst

TABLE XIV

High-Surface-Area Catalysts Performance[a]

Catalyst	Activity (g PP/g catalyst)	Isotactic index
$3TiCl_3 \cdot AlCl_3$[b]	942	89.4
Shell type[c]	2012	95.2
Solvay type[d]	4028	93.5

[a] Polymerization conditions: 142 psig C_3, 3.2 psig H_2, 70°C, 3 hr; DEAC:Ti mol ratio 3.3, n-heptane solvent.
[b] Stauffer $TiCl_3$ 1.1.
[c] Prepared according to Kortbeek et al. [21].
[d] Prepared according to Hermans and Henrioulle [20].

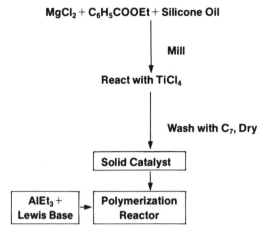

MgCl$_2$ + C$_6$H$_5$COOEt + Silicone Oil

↓ **Mill**

React with TiCl$_4$

↓ **Wash with C$_7$, Dry**

Solid Catalyst

↓

AlEt$_3$ + | **Polymerization**
Lewis Base → **Reactor**

Fig. 6. Supported titanium polypropylene catalyst. (From Luciani *et al.* [22].)

development continues today with the goal of producing a catalyst having a low titanium level but an activity and stereospecificity high enough to avoid both deashing and removal of atactic polymer.

In a typical supported catalyst prepared as illustrated in Fig. 6, the titanium level can range from 1 to 4%. If 3% titanium is present in the catalyst, an activity of over 6000 g of polypropylene per gram of catalyst (or 200,000 g of polypropylene per gram of titanium) will result in 5 ppm of titanium or less in the polymer. After combining research efforts, in 1976 Montedison and Mitsui announced that an activity of 300,000 g of polypropylene per gram of titanium had been achieved in combination with an isotactic content of 95%. Since then, announcements have claimed use of this new technology in plants in Italy, Belgium, Austria, and the United States. Reportedly, a savings of 1–2¢/lb of polymer in manufacturing costs and a 10–20% savings in capital investment for new plants can be realized with this catalyst.

In most examples cited in the patent literature, the support material is a magnesium compound. Anhydrous magnesium chloride appears to be particularly effective for both activity and stereospecificity. Physical properties that may contribute to this include the similarity of crystal structures of MgCl$_2$ and γ-TiCl$_3$ and the nearly identical ionic radii of Ti^{4+} and Mg^{2+}, which are 0.68 and 0.65 Å. These characteristics permit the titanium to occupy the defects in the MgCl$_2$ lattice generated through grinding or solvent etching. Then, on reduction by aluminum trialkyl, the titanium halide formed is in a steric environment very similar to γ-TiCl$_3$. As shown in Fig. 6, in practice the titanium compound is complexed with an electron donor to improve the stereospecificity of the catalyst. The data in Table XV demonstrate that not all electron donors are effective in promoting the

202   Kelly B. Triplett

Effect of Lewis Bases Complexed to TiCl₄ in High-Mileage Catalysts[a,b]

Ti catalyst	Polypropylene yield [(g/g Ti/hr/atm) × 10⁻³]	Isotactic index
MgCl$_2$:0.15 TiCl$_4$	2.4	51
MgCl$_2$:0.15 TiCl$_4$:0.7 TMB	7.2	55
MgCl$_2$:0.15 (TiCl$_4$ × POCl$_3$):0.7 TMB	1.5	51
MgCl$_2$:0.15 (TiCl$_4$ × PhCO$_2$Et):0.7 TMB	2.1	80
TiCl$_4$ × PhCO$_2$Et, unmilled	0.02	71
TiCl$_4$	0.35	26

[a] Reprinted with permission from Duck, E. W.; Grant, D.; and Kronfli, E., *Eur. Poly. J. 15*, 626, (1979), Pergamon Press, Ltd. [23].

[b] Reaction conditions: 1 atm total pressure, glass reactors, stirred at 800 rpm, Et$_3$Al 5 mmol, Ti catalyst 0.03 mmol, and 500 ml isooctane as solvent. TMB, 1,2,4,5-Tetramethylbenzene.

formation of isotactic polypropylene. Aromatic esters, such as ethyl benzoate, produce some of the more improved levels of isotactic polymer.

When an aluminum trialkyl alone is used as the cocatalyst in the polymerization reactor, an electron donor complex of TiCl$_4$ is insufficient for providing an isotactic index much above 80. To increase this value above 90, a second electron donor is added *in situ* with the aluminum alkyl. The molar ratio of electron donor to aluminum alkyl is critical to catalyst performance. The data in Table XVI indicate that the activity is inversely proportional, and the isotactic index directly proportional, to this molar ratio. Pino and Mülhaupt in 1980 [24] showed that, with an increasing electron donor/aluminum alkyl ratio, atactic formation decreased appreciably, whereas isotactic polymer production decreased by only ~ 25%. The

Influence of Aluminum Alkyl: Electron Donor Ratio on Supported Catalysts[a]

Et$_3$Al/ethyl anisate mole ratio	Activity (g PP/g catalyst)[b]	Isotactic index
2.7	719	94.7
3.3	960	90.2
3.7	1138	88.4
4.0	2054	88.6
5.0	2515	84.2

[a] Polymerization conditions: 142 psig C$_3$, no H$_2$, 65°C, 1.5 hr, *n*-heptane solvent. PP, Polypropylene.

[b] Catalyst prepared by grinding MgCl$_2$ plus TiCl$_4$ · ethyl benzoate.

TABLE XVII

Supported Catalyst Comparison of Slurry versus Bulk
Polymerization

Polymerization type	Activity (g PP/g catalyst)[a]	Isotactic index
Slurry[b]	7,858	94.1
Bulk[c]	12,311	96.1

[a] $MgCl_2$ plus ethyl benzoate plus silicon oil milled, plus reaction with $TiCl_4$.
[b] Polymerization conditions: 142 psig C_3, no H_2, 65°C, 1.5 hr, n-heptane solvent, Triethylaluminum plus ethyl anisate cocatalyst.
[c] Polymerization conditions: 470 psig C_3, no H_2, 70°C, 1.5 hr, Triethylaluminum plus ethyl anisate cocatalyst.

net result was an increase in isotactic level but an overall decrease in activity.

This behavior was explained by competition for the electron donor between the aluminum alkyl and active sites of varying stereoregulating capacity. If the sites producing atactic polymer have higher Lewis acidity, they would be more easily deactivated by an electron donor. In many instances, this second electron donor is again an aromatic ester, with ethyl p-anisate being frequently cited.

Supported catalysts appear to be particularly suited to the bulk polymerization process where a high monomer/catalyst concentration ratio exists. This is advantageous in situations where the polymerization rate decreases rapidly with time, as is the case with many supported Ziegler–Natta catalysts. However, good performance is also found in slurry polymerization processes, as indicated in Table XVII. Despite the dramatic improvement in performance in terms of grams of polypropylene per gram of titanium, some technical problems still exist. Because of the corrosive nature of $MgCl_2$ in extruders and other processing equipment, neutralization agents must be added to the polymer if deashing is to be completely eliminated. Implementation of the supported catalyst in a polymerization plant requires new technology because of extremely high polymerization rates and the critical role of the correct ratios of components. In fact, it has been reported that the supported catalyst provides no substantial advantages for the bulk loop reactor [13]. On the other hand, Rexene claims to have successfully introduced a supported catalyst into their bulk polymerization process.

VII. Future Developments

In the future, research and development in Ziegler–Natta catalyst technology will probably continue at the same high pace as that seen during the past

25 yr. The majority of this activity will be concentrated in the supported catalyst area in application of the catalyst for block copolymerization; reduction of corrosive residuals; formation of uniform, spherical particles to eliminate pellet production; and in application to gas phase polymerization reactors. Further modifications in the catalyst, cocatalyst, and polymerization conditions are required, since the resulting unextracted polymer does not always exhibit the same molding characteristics found for extracted polypropylene produced from conventional catalysts.

Ziegler–Natta catalyst development will enter its next stage upon the introduction of homogeneous systems for the production of isotactic polypropylene. Early work on homogeneous systems was not successful in offering a commercially acceptable catalyst system. However, a homogeneous system with the activity and isotactic index at supported catalyst levels still would be very attractive commercially. Many problems inherent in a heterogeneous system, such as nonuniformity of performance, particle attrition, and equipment for accurate metering into the reactor, could be avoided. Theoretically, with proper design a homogeneous system should be effective in producing isotactic polymer at the levels needed commercially. When this development occurs, the evolution of Ziegler–Natta catalysts will have reached a new pinnacle.

References

1. K. Ziegler, BE Patent 533, 362 (German priority 1953).
2. K. Ziegler, E. Holzkamp, H. Breil, and H. Martin, *Angew. Chem.* **67**, 541–547 (1955).
3. G. Natta, *J. Polym. Sci.* **16**, 143–154 (1955).
4. J. Boor, "Ziegler–Natta Catalysts And Polymerizations." Academic Press, New York (1979).
5. S. Gross, (ed.). *Mod. Plast.* **59**, 81 (1982).
6. G. Natta, I. Pasquon, A. Zambelli, and G. Gatti, *Makromol. Chem.* **70**, 191–205 (1964).
7. R. F. Gold, S. F. Gelman, J. R. Doonan, and G. G. Arzoumanidis, *Chem. Eng. (N.Y.)* **84** No. 3, 119–121 (1977).
8. G. Natta, *Chim. Ind. (Milan)* **41**(6), 519 (1959).
9. P. Parrini, F. Sebastiano, and G. Messina, *Makromol. Chem.* **38**, 27–38 (1960).
10. G. Natta, *J. Polym. Sci.* **34**, 21–48 (1959).
11. G. Natta, P. Pino, G. Mazzanti, U. Giannini, E. Mantica, and M. Peraldo, *J. Polym. Sci.* **26**, 120–123 (1957).
12. G. Natta and G. Mazzanti, *Tetrahedron* **8**, 86–100 (1960).
13. J. N. Short, *CHEMTECH* **11** No. 4,238–243 (1981).
14. W. Klemm and E. Krose, *Z. Anorg. Chem.* **253**, 209–217 (1947).
15. G. Natta, P. Corradini, I. Bassi, and L. Porri, *Rep. Ital. Nat. Acad. Sci. Ser.* **24**, book 2 (1958).
16. E. Tornquist and A. W. Langer, U.S. Patent 3,032,510 Esso Research and Engineering Company (1962).

17. G. Natta, P. Corradini, and G. Allegra, *J. Polym. Sci.* **51**, 399–410 (1961).
18. P. Cossee, *J. Catal.* **3**, 80–88 (1964).
19. G. G. Arzoumanidis, U.S. Patent 4,124,530, Stauffer Chemical Company (1978).
20. J. P. Hermans and P. Henrioulle, U.S. Patent 4,210,738, Solvay and Cie (1980).
21. A. Kortbeek, A. van der Nat, W. van der Linden Lemmers, and W. Sjardijn, U.S. Patent 4,195,069, Shell Oil Company (1980).
22. L. Luciani, N. Kashiwa, P. C. Barbe, and A. Toyota, DT patent 2,643,143, Montedison and Mitsui (1977).
23. E. W. Duck, D. Grant, and E. Kronfli, *Eur. Polym. J.* **15**, 625–626 (1979).
24. P. Pino and R. Mülhaupt, *Angew. Chem. Int. Ed. Engl.* **19**, 857–875 (1980).

CHAPTER 8

Ethylene Oxide Synthesis

J. M. BERTY*
Berty Reaction Engineers, LTD.
Erie, Pennsylvania

I. Ethylene Oxide

Ethylene oxide, one of the most versatile chemical intermediates, is currently produced by catalytic oxidation of ethylene almost exclusively. The sales value of the ethylene oxide produced in the United States exceeds $1 billion. The high chemical reactivity of ethylene oxide gives it its desirable qualities as a chemical intermediate and also its dangerous properties.

A. PHYSICAL PROPERTIES

At room temperature and pressure ethylene oxide is a colorless gas that is explosive even at 100% concentration, i.e., without any admixture of air or

---

* Present address: Department of Chemical Engineering, University of Akron, Akron, Ohio, 44325.

oxygen. At low temperatures it condenses to a mobile liquid. It is miscible with water, alcohol, ether, and many organic solvents in all proportions. Most of its interesting properties result from the strained ring structure. Structural studies were reviewed by Parker and Isaacs [1]. More information can be found in the paper by Fujimoto [2] and the book by Coulson [3].

The most important physical properties of ethylene oxide are given in Tables I–III. Additional information on physical properties can be found in Yaws and Rackely [4] and in L'Air Liquide [5].

B. CHEMICAL PROPERTIES

Ethylene oxide is a valuable intermediate in preparing a variety of compounds because of its high reactivity. The strained ring structure opens up easily and attaches to most compounds that have an active hydrogen in the molecule, thereby introducing a hydroxyethyl group:

$$RH + CH_2 - CH_2 \overset{O}{\diagdown} \longrightarrow R(CH_2CH_2O)H$$

The hydrogen on the terminal hydoxy group is also reactive, and therefore it can further react with ethylene oxide:

$$RCH_2CH_2OH + nCH_2 - CH_2 \overset{O}{\diagdown} \longrightarrow R(CH_2CH_2O)_nCH_2CH_2OH$$

With *water* ethylene oxide forms glycol and polyglycols. The reaction proceeds well at an elevated temperature and pressure and also with the help of acid or base catalysts. In the industrial production of glycol, a large excess of water is used to minimize the formation of di- and higher polyglycols, because ethylene glycol is the most valuable product (per unit of ethylene oxide used) and is required in the largest quantities. The ethylene glycol produced is therefore in the form of a dilute water solution and needs to be concentrated and refined to a saleable product.

$$HOH + CH_2 - CH_2 \overset{O}{\diagdown} \longrightarrow HOCH_2CH_2OH$$

$$HOCH_2CH_2OH + CH_2 - CH_2 \overset{O}{\diagdown} \longrightarrow HOCH_2CH_2OCH_2CH_2OH$$

With *alcohols* the reaction is similar to that with water, a monoether of ethylene glycol being the primary product. This compound can then further react with ethylene oxide, forming monoethers of polyethylene glycols. These products are water-soluble to a great degree, depending on the molecular weight and the length of the alcohol moiety.

TABLE I

Physical Constants of Ethylene Oxide[a]

Property	Value
Molecular weight	44.05
Boiling point (°C)	
At 101.3 kPa (760 mmHg)	10.4
Boiling point/pressure at 100 kPa (K/kPa, K/mmHg)	0.25 (0.033)
Freezing point (°C)	−112.6
Critical pressure (MPa, atm)	7.19 (71.0)
Critical temperature (°C)	195.8
Explosive limits in air (vol%)	
Upper	100
Lower	3
Flash point, tag open cup (°C)	−18
Heat of combustion at 25°C (kJ/mol)	1306.04
Heat of fusion (kJ/mol)	5.17
Heat of solution in pure water at 25°C and constant pressure (kJ/mol)	6.3
Health hazard threshold limit (1971) (ppm)	50

[a] From Kirk–Othmer [10]. Data from Union Carbide (1973), Walters and Smith [91], Pell and Pichler [92], and Giauque and Gordon [93].

With *ammonia and amines* ethylene oxide can react occupying the free hydrogen atoms on the nitrogen. These atoms are more reactive than the hydrogen in the hydroxyl group, yet the presence of a small quantity of water is essential for the reaction. Ethanolamines are important solvents and chemical intermediates.

The *polymerization* of ethylene oxide to low-molecular-weight polymers (oligomers, MW 200–10,000) can be started with water and alcohols if ethylene oxide is added to the primary product. Alkali alcoholates are the most used catalysts [6]. Higher polymers with molecular weights up to 10^6 are formed by anionic polymerization [7]. The nonvolatile residue in ethylene oxide storage tanks is a polymer formed on iron rust [8].

The *isomerization* of ethylene oxide to acetaldehyde is not an industrial reaction but is important in the catalytic reactor operation. It is highly exothermic, and several surfaces can catalyze it.

$$C_2H_4O \rightarrow CH_3CHO$$

$$\Delta H° = -115 \ \text{kJ/mol}, \qquad \Delta G° = -122 \ \text{kJ/mol}$$

The *decomposition* of ethylene oxide can be explosive, even that of the 100% vapor. To set off a 100% ethylene oxide vapor detonation the energy required is four orders of magnitude larger than the energy input needed for

TABLE II

Thermodynamic and Transport Properties of Ethylene Oxide[a]

Temperature (K)	Entropy (J/mol K)	Heat of formation (kJ/mol)	Free energy of formation (kJ/mol)	Viscosity (Pa sec)	Thermal conductivity (W/m K)	Heat capacity (J/mol K)
298	242.4	−52.63	−13.10	—	0.012	48.28
300	242.8	−52.72	−12.84	9.0	0.025	48.53
400	258.7	−56.53	1.05	13.5	0.038	61.71
500	274.0	−59.62	15.82	15.4	0.056	75.44
600	288.8	−62.13	31.13	18.2	0.075	86.27
700	302.8	−64.10	46.86	20.9	0.090	95.31
800	316.0	−65.61	62.80	—	—	102.9

[a] From Kirk–Othmer [10]. Data from Pell and Pilcher [92], Gallant [94], Yaws [95], Mass and Boomer [96], Stull et al. [97], Kobe and Pennington [98].

TABLE III

Physical Properties of Aqueous Solutions of Ethylene Oxide[a]

Weight percent	Mole percent	Specific gravity at 15/15°C	Freezing point (°C)	Boiling point (°C)
0	0	1.000	0.0	100
2.5	1.0	0.9993	−0.9	70
5	2.1	0.9986	−1.6 (eutectic)	58
10	4.4	0.9973	5.6	42.5
15	6.7	0.9959	8.9	38
20	9.3	—	10.4	32
30	14.9	—	11.1 (max)	27
40	21.4	—	10.4	21
50	29.0	—	9.3	19
60	38.0	—	7.8	16
70	48.8	—	6.0	15
80	62.1	—	3.7	13
90	78.6	—	0.0	12
100	100	—	−112.5	10.4

[a] From Kirk–Othmer [10]. Data from Union Carbide [9] and Mass and Boomer [96].

ethylene oxide–air mixtures. The decomposition reaction is a chain reaction that proceeds through acetaldehyde, and the products are the usual decomposition compounds. An approximate idea of the energy release can be gained by considering the following oversimplified example:

$$C_2H_4O \rightarrow CH_4 + CO$$

$$\Delta H° = -134.2 \quad kJ/mol, \quad \Delta G° = -176.2 \quad kJ/mol$$

Theoretically, the decomposition pressure of ethylene oxide vapor can exceed the initial pressure by more than 10 times if decompostion starts at 100°C. Although liquid ethylene oxide cannot be detonated, explosion of the vapor space above it can evaporate more liquid and make an accident worse. A strong igniter, such as a detonation cap or a melting 30-gauge Nichrome wire, can initiate a deflagration (an explosion but not a detonation) in liquid ethylene oxide, and the consequences are more dangerous; because of the greater concentration of material stored the potential energy is larger.

The vapor space over stored liquid ethylene oxide should be blanketed (diluted) by inert gas to keep the vapor phase concentration below the explosive limit. In case of a spill, ethylene oxide should be diluted with 23 volumes of water to keep it below the flash point.

TABLE IV

Technologies and Production Capacities

	Capacity (t/yr)	
Technology	1964	1978
Union Carbide	636	1134
Shell	245	732
Scientific Design	195	533
Dow	91	291
	1167	2690

C. PRODUCTION STATISTICS

In the United States and Puerto Rico 12 companies produce ethylene oxide, using 4 different technologies, at 15 locations. The United States provides more than half of the world's production.

Table IV summarizes the capacities and technologies used in 1964 and 1978 from the detailed statistics presented in the second and third editions of Kirk–Othmer [9, 10].

Union Carbide, the leading producer, in 1964 had production plants at eight different locations. In 1980 it had plants at only three locations, but these three had almost twice the capacity of the former eight. The smallest capacity Union Carbide plant in Puerto Rico is larger than the largest of all the other producers. Union Carbide and Dow use their own technology, Shell technology is used in six plants, and Scientific Design's in four.

Production in the United States in 1980 was 2250 t, and in 1979 it was 2580 t. From a 1979 high production fell about 13% to the 1980 level (*C & EN*, June 8, 1981).

The price of ethylene oxide was 39–40½ ¢/lb in 1980 and 45 ¢/lb in 1981. Ethylene prices in the same years were 22–24 and 28 ¢/lb, respectively [11, 12]. 1982 and 1983 prices have declined.

II. Synthesis

A. HISTORIC BACKGROUND

Ethlyene oxide was first prepared more than a century ago. Wurtz [13] obtained it from ethylene chlorohydrin but failed to make it by oxidation [14]. Lefort [15] succeeded in the air-oxidation of ethylene to ethylene oxide. Union Carbide Corporation, then called Carbide and Carbon Chemicals Corporations, purchased the Lefort patent and started to develop a commercial process. The first commercial production was started at Union Carbide Corporation in 1937. Oxygen-based direct oxidation was started by Shell in 1958.

The former I. G. Farbenindustrie wanted to buy the Lefort patent in 1932 [16], but the high price and the lack of developmental results discouraged it. Later I. G. Farbenindustrie developed its own process, yet it bought the Lefort patent in 1938 from the Société Française de Catalyse General and built a 1100-lb/day pilot plant for the process. In England, Distillers Company constructed a similar experimental unit. The first larger scale plant was built in Germany toward the end of World War II in 1943 in Zweckel with a capacity of about 25 million lb/year, but this unit was never started up [16].

B. CHLOROHYDRIN PROCESS

The first commercial route to the production of ethylene oxide was the chlorohydrin process. Here, ethylene first reacts with hypochlorous acid to form ethylene chlorohydrin. This compound, in a second step, reacts with lime to produce ethylene oxide and a by-product, calcium chloride.

Badische Anilin- & Soda-Fabrik (BASF) started commercial production by this route during World War I in Germany. In the United States, Union Carbide Corporation started using it in 1925. With the introduction of the direct oxidation process, these plants were shut down or utilized for propylene oxide manufacture from propylene by the chlorohydrin process.

At present only one large unit is known to be using this method: Dow Chemical Company's plant at Freeport, Texas.

III. Ethylene Oxidation

A. NONCATALYTIC OXIDATION OF ETHYLENE

In the early experiments on ethylene oxidation only formaldehyde was observed as a product. Schutzenberger [17] discovered that at about 400°C oxidation started and formaldehyde was formed in small quantities. Willstatter and Bommer [18] studied the reaction first in relation to the manufacture of formaldehyde by both catalyzed and noncatalytic routes. With a little ethylene in air at 600°C and a short reaction time, 3% of the introduced ethylene was converted and 50% of it became formaldehyde.

Even more recently the economic incentive to use ethane or ethane – ethylene mixtures, instead of the more expensive pure ethylene, was strong enough to justify the development of a noncatalytic process for making ethylene oxide [19]. Although the production of propylene oxide is possible from propane–propylene mixtures with over 20% yield coproduced. An additional 30% useful products such as acrolein and acetaldehyde are the oxidation of ethylene–ethane mixtures has been less efficient in ethylene oxide formation.

B. CATALYTIC OXIDATION OF ETHYLENE TO ETHYLENE OXIDE

1. Properties of Silver

Silver is the unique catalyst for ethylene oxidation. All practical catalysts are silver-based. Silver is the best electrical conducting material, with a conductivity of 1.67 $\mu\Omega$/cm. It is the second best conductor of heat after diamonds, with a thermal conductivity of 4.29 W/cm K.

Although there are conflicting data on the adsorption of ethylene, ethylene oxide, water, and carbon dioxide on pure metallic silver, in view of the extreme difficulty of preparing clean silver surfaces, it seems safe to say that none of these compounds is adsorbed to any significant extent. Ethylene oxide primarily, and carbon dioxide to a much lesser extent, may be adsorbed and then quickly react and decompose on silver surfaces, contaminating the surface with oxygenated species. The difficulty of obtaining clean and reproducible surfaces was shown by Czanderna [20] and by others.

2. The Silver–Oxygen System

Silver oxide decomposes above room temperature in air. For the reaction

$$2Ag + \tfrac{1}{2}O_2 \rightarrow Ag_2O,$$

$$\Delta H° = -28 \quad kJ/mol, \qquad \Delta G° = -10 \quad kJ/mol.$$

The decomposition pressure is a function of temperature. According to Dushman [21] silver oxide needs 0.56 and 20.8 atm of oxygen pressure to remain stable at 173 and 302°C, respectively, where the oxidation reaction occurs. Although silver oxide is not stable at the temperature of ethylene oxidation, the surface of silver readily adsorbs oxygen at these temperatures.

Ostrovskii and Temkin [22] measured the heat of adsorption of oxygen on silver and found that it dropped from 500 kJ/mol at low coverage to about 42 kJ/mol of O_2 at full coverage; therefore at low coverages the heat of adsorption is higher than the heat of formation for silver oxide. The curve for the heat of adsorption versus the surface coverage shows two breaks,

TABLE V

Activation Energies of Absorption and Desorption and Heat of Adsorption of Oxygen on Silver[a]

Activation energy		Heat of absorption (kJ/mol)	Reference
Adsorption (kJ/mol)	Desorption (kJ/mol)		
39–49	—	—	Meisenheimer and Wilson [66]
92	—	—	Smeltzer et al. [100]
13–34–92	—	105–113	Czanderna [20, 43]
	136	67	Sandler et al. [101]
	71	—	Sandler et al. [102]
13–34–59	142	—	Kilty et al. [52]
	159	—	Kollen and Czanderna [103]
	147–188	—	Rovida et al. [104]
	67–117	—	Czanderna et al. [105]
		67	Benton and Drake [106]
53		75–80	McCarty [107]
34–42		71–105	Smeltzer et al. [100]
		67	Sandler et al.
		452–544	Ostrovskii and Temkin [22]
		—	Kulkova and Temkin [108]
62		—	Clarkson and Cirillo [109]

[a] From Verykios et al. [27].

suggesting three different adsorbed oxygen species. Some of the oxygen can be dissolved in the solid silver also. Czanderna [23, 24] made a comprehensive study on the rate of adsorption and desorption of oxygen on silver in the range -77 to $351°C$ using a microbalance technique. The measured energies of activation in three different regions were 13, 34, and 92 kJ/mol. These values were considered to represent the adsorption of O^- and O_2^- and the mobility of O_{ads}. At the higher end of the temperature range charged atoms and charged molecules dominated.

Sachtler [25] questioned the validity of Czanderna's proof of the existence of molecular, i.e., diatomic, oxygen at the surface, yet in a later work with Janssen *et al.* [26] showed, using field emission microscopy, that both atomic and molecular oxygen existed on the silver surface at very low pressures. Other authors came to the same conclusion and, in spite of the continued controversy about certain details, it is safe to assume that molecular oxygen exists on the surface, and this is important in considering the mechanism of ethylene oxide formation over a silver catalyst.

A recent review of the general field of ethylene oxidation that lists the large number of works on chemisorption studies was made by Verykios *et al.* [27]. In most published studies, because of the applied instrumentation, either very low pressures or a vacuum, usually low temperatures, or both were used, thereby making the studies far from representative of the conditions used in commercial production. Although all these studies have contributed very significantly to our understanding of the silver–oxygen system and its interaction with ethylene and the products, one has to be careful in using these results to explain the mechanism of the oxidation process operating under quite different conditions.

Oxygenated silver adsorbs ethylene and ethylene oxide, as well as carbon dioxide and water. Marcinkowsky and Berty [28] showed that ethylene is adsorbed on oxygenated silver in part reversibly and in part irreversibly. The measurements were made by a frontal chromatographic technique at atmospheric pressure and near reaction temperature. The amount of reversibly adsorbed ethylene decreased with increasing temperature, whereas the amount of irreversibly held ethylene increased. This irreversibly held ethylene was considered chemically adsorbed. Force and Bell [29] have shown by infrared spectroscopy that ethylene adsorbs on oxygenated silver without rupture of the carbon–carbon double bond. Force and Bell have also [30] reported that CO_2 adsorbs on oxygenate silver in three different forms and have presumed that water adsorbs dissociatively, forming two hydroxyl groups. Evidence for water adsorption on a partially oxidized surface was reported by Benton and Elgin [31] a long time ago, but not much has been done since, perhaps because of the difficulties of handling water in a high-vacuum system.

3. Ethylene Oxidation with Pure Silver

Most of the research work on ethylene oxidation was limited to the study of pure systems consisting of ethylene, oxygen, and silver only. It is questionable, especially in the early work, how pure these systems were, especially in view of the difficulty many investigators had in obtaining reproducible results and the repeated emphasis in later work on the effects of small contaminations.

Pure silver is not very stable in finely dispersed form or even as shiny, low-surface-area metallic foil. Although high-surface-area pure silver powder sinters during oxidation, if silver metal foil is exposed to alternating oxygen and ethylene at the usual reaction temperature, it becomes grayish in color, the surface area grows, and eventually it starts to produce small quantities of ethylene oxide. Yield, selectivity, and productivity are all low for pure silver, and performance declines rapidly because of agglomeration of the silver particles.

Early published theories on ethylene oxide formation on silver catalyst were advanced by Twigg [32] and in less well publicized work by Worbs [33], Meter [34], and Schultze and Thiele [35]. The first assumed a reaction between atomically adsorbed oxygen and ethylene, the others assumed reactive, peroxide-type O_2 on the surface in explaining the epoxidation. The second theory involving peroxide-type oxygen is the one most accepted today. This, in the view of the author, is also supported by the fact that in organic chemical preparations of epoxides by oxidation in the liquid phase, as well as in high-temperature gas phase reactions, epoxides are usually produced from olefins by the action of peroxides. The reason for the uniqueness of silver in catalyzing ethylene oxide formation is not known definitely, but its ability to form peroxide-type O_2 species on the surface under the required reaction conditions is part of the explanation.

From the peroxidic oxygen mechanism for ethylene oxide formation some conclusions can be drawn regarding the possible maximum selectivity. For example:

$$Ag + O_2 \rightarrow Ag \cdot O_{2,ads}$$

$$Ag \cdot O_{2,ads} + C_2H_4 \rightarrow Ag \cdot O_{ads} + C_2H_4O$$

If the above reactions occur six times, then a seventh ethylene molecule will be needed to clean off the remaining atomic adsorbed oxygen to free the silver surface to accept molecular oxygen again.

$$6Ag \cdot O_{ads} + C_2H_4 \rightarrow 2CO_2 + 2H_2O + 6Ag$$

This theory assumes, from bond energy balances, that oxygen can not regenerate the $Ag \cdot O_{ads}$ to $Ag \cdot O_{2,ads}$ from the gas phase. The maximum

TABLE VI

Activation Energies for the Partial and Complete Oxidation Reaction
of Ethylene over Silver Catalysts[a]

Catalysts	Activation energy (kJ/mol)		References
	To C_2H_4O	To $CO_2 + H_2O$	
Ag, not described	63	88	Kurilenko et al. [110]
Ag, Ba-promoted	50	63	Murray [111]
Ag, on alumina, Ba-promoted	80	—	Wan [112]
Ag, promoted	80	90	Ostrovskii et al. [51]
Ag, evaporated film	109	124	Twigg [32]
Ag, on alumina	90	121	Kenson and Lapkin [113]
Ag, on glass wool	96	—	Twigg [32]
Ag, on corundum	90	142	Boreskov et al. [76]
Silver oxide	63	57	Liberti et al. [114]
Ag, thoroughly cleaned	—	88	Korchak and Tret'yakov [115]

[a] From Verykios et al. [27].

possible selectivity of this mechanism is $\frac{6}{7}$, or 85.7%. Other similar theories give 80% or lower selectivities, but these levels have been exceeded in reliable experiments, hence no real support exists for them [36]. Voge and Adams [37] stated that there is a "bewildering amount of information on the kinetics of ethylene oxidation on silver" and presented a good overview of the results up to 1967.

In addition to the primary parallel formation of CO_2 with ethylene oxide, some CO_2 is also formed from the secondary oxidation of ethylene oxide. The general, triangular scheme is

$$C_2H_4 \xrightarrow{r_1} C_2H_4O$$
$$r_2 \searrow \swarrow r_3$$
$$CO_2$$

On a good industrial catalyst the ratio of the rates can be $r_1/r_2 \simeq 6$ and $r_2/r_3 \simeq 2.5$. These values are for orientation only, because they depend strongly on the reaction conditions in addition to the catalyst.

Several authors have found that the selectivity hardly changes with increasing conversion until conversion becomes high. This is observed in laboratory experiments, where rates are low, and also in commercial converters that have high heat transfer rates and good temperature control. Secondary oxidation kinetics can be studied in the laboratory best by using isotope-labeled ethylene oxide in the feed.

The secondary oxidation of ethylene oxide is a catalytic process also, although some possibility exists that it proceeds through the heterogeneous-homogeneous route [38]. This may be more important in production reactors, operating at a high temperature and high heat release rate, where some parts of the catalyst are overheated and they become free radical sources. Some differences in the selectivity of the air and oxygen processes may be associated with differences in the ability of nitrogen and hydrocarbon gases to "adsorb" free radicals [39], i.e., to terminate radical chains. Although this route is not proven, it remains a possibility in the opinion of the author that some free radical mechanism is involved in the temperature runaway phenomenon that starts on an overheated catalyst. It can be speculated that the source of free radicals on an overheated catalyst is the decomposition of a peroxide-type compound. This may be absorbed molecular oxygen and, as in other peroxy compound decompositions, the desorbed oxygen is in a reactive form like singlet oxygen [40]:

$$Ag-O-O \rightarrow Ag + O_2(^1\Delta g)$$

Another possibility is that the intermediate complex that supplies the ethylene oxide splits at high temperature (above 320°C) not at the oxygen–oxygen bond but at the silver–oxygen bond:

$$
\begin{array}{c}
Ag-O-O \overset{CH_2}{\underset{CH_2}{\diagup}} \\
\end{array}
$$

$Ag \!-\! O \!+\! O \overset{\diagup CH_2}{\underset{\diagdown CH_2}{}}$

$$T > 320°C \qquad\qquad T > 320°C$$

$Ag + O\!-\!O \overset{\diagup CH_2}{\underset{\diagdown CH_2}{}} \qquad\qquad Ag\!-\!O + O \overset{\diagup CH_2}{\underset{\diagdown CH_2}{}}$

The temperature rise during temperature runaways, as indicated by thermocouples in the gas phase, is so fast that it cannot occur on the catalyst, where mass transfer limits the reactant supply and heat transfer limits the heat release rate. Although the importance of the several surface oxygen species is generally accepted, there are theories other than that involving the reaction of peroxide-type oxygen with gaseous ethylene that cannot be discarded completely. The suggestion that subsurface oxygen converts Ag to Ag$^+$ [41], which in turn can absorb ethylene, is not totally contradictory to the peroxide mechanism, and it is similar to a previous proposal by Temkin [42].

Since 1967 information on kinetics more than doubled in volume, yet no common organizing idea evolved that could explain all the observed facts. Some of the possible reasons for the large divergence of results may be related to the facts that the catalyst itself is unstable and undergoes changes during the reaction and that these changes are slow.

Czanderna stated [43] that silver, carefully stabilized by repeated oxygen adsorption and O_2 removal by reduction with CO, increased its surface area by about 10% when very little ethylene was added at a very low pressure and that a reaction occurred.

Maxwell [44] stated that the size of silver particles changes during ethylene oxidation. Particles larger than 4 μm break up, and those smaller than 0.1 μm sinter and form more stable particles. Since all these changes are slow, steady operation for hours, days, and sometimes weeks is needed for the catalyst to come to a state of dynamic equilibrium with the reacting media and with the highly exothermic reaction proceeding on the surface. Consequently, investigators who do not have the time to wait out these slow, transient changes in the solid phase obtain very interesting results that add to the confusion. Also, most physical investigations of catalysts are made on new, unused catalyst samples and not on properly broken-in catalysts that have come to a steady state with the reacting mixture. Real progress is made only slowly, where funds and time permit expensive, time-consuming studies, e.g., at large companies who in turn do not publish much in detail.

C. CATALYST PREPARATION

1. Patent Literature

As is usual in the patent literature, anything and everything is patented for ethylene oxidation. One fact stands out in all these publications, and this is the significance of silver. No catalyst is known that can produce ethylene oxide in good yield and high productivity that does not contain silver. Almost all of the other elements in the periodic table are claimed to be useful additives for one purpose or another. From these the significance of earth alkali additives emerges as the most important, followed by alkali metals among the cations [45]. Of the anion-type constituents chloride seems to be the most important. Most of these components have an optimum concentration as additives, meaning primarily that both good and bad effects are expected. Some, like heavy metals such as copper, have very deleterious effects at very small concentrations; others, like potassium, have good effects but only at very small concentrations. The bad effect of excess additives was noted by Bathory et al. [46] and by Carberry et al. [47] and dramatically documented in a patent by Bhasin et al. [48].

2. Role of Earth Alkali Additives

This subject is still very controversial, perhaps because more than one role can be attributed to these additives, and various investigators concentrate on one or another of the effects in their studies. One effect is recognized by Spath et al. [49] as stabiliation of the catalyst matrix by preventing sintering. Another effect of covering the silver with BaO_2 or $BaCO_3$ is the creation of a semiconducting layer and a decrease in the work function of the catalyst matrix. A similar theory was proposed by Margolis in 1963 [38], yet the significance of the electronic effects with silver, the best electrical conductor of all materials, is not generally accepted. Perhaps viewing the oxygenated silver – earth alkali mixture as a semiconducting catalyst, with the metallic silver only as a support, would give a better picture. Another effect of early alkali and alkali additives on silver is the removal of anionic contaminants, e.g., chloride, sulfide, and sulfate, as will be shown later. Yet another overall effect is binding the silver mixture to the catalyst carrier by creating better adhesion. $BaSiO_3$ is used in some cracking catalysts as a binder.

Earth alkali additives are used in 1 – 15 wt% quantities based on silver, while alkalis are used in the 0.005 – 0.05 wt% range also based on silver content.

3. Anionic Part of Additives

The addition of halogenated organic compounds in very small quantities to the feed gas considerably improves the selectivity. This fact is known from the patent obtained by Law and Chitwood [50]. The effect of the addition of halogens to the catalyst at preparation was studied by Ostrovskii et al. [51]. In addition to chlorine, sulfur, selenium, and tellurium increased the activity in very small quantities. Further increases in these compounds decreased activity but increased selectivity. These effects were explained by changes in the bond energies of the adsorbed oxygen resulting from the presence of these additives. Kilty et al. [52], starting with the view that molecular oxygen produces the ethylene oxide and atomic O_2 the by-product, theorize that chlorine inhibits the dissociative adsorption of oxygen.

In the patent literature Sears [53] claims that alkali metal halides can be advantageous at 0.01 to 50.0% based on the weight of the metallic silver. The halide content was claimed to enhance the break-in period and maintain moderate but steady activity over an extended period. The effects of halides of different metals vary. In a new patent Bhasin et al. [48] claims that metal halides built into the catalyst inhibit the total combustion of ethylene and classifies them as depressants or anticatalysts.

4. Catalyst Supports

The active ingredient of the catalyst has to be distributed on a support, primarily for economic reasons, to obtain a better performance per pound of silver used. Another reason is to limit the heat generation rate per unit volume. The author estimates that U.S. ethylene oxide plants contain 10–20 million troy oz of silver in the form of catalysts.

The most important requirement of the catalyst carrier or support is complete inertness. The preferred support seems to be α-alumina. Other frequently mentioned carrier materials are carborundum and silica. Thermal conductivity, often considered important, is better for carborundum than for alumina, yet alumina is usually the choice. The reason may be that at the high mass velocities of a commercial operation convective heat transfer dominates and the thermal conductivity of the carrier loses its importance.

The required surface area for the support is very low in view of the large surface areas associated with most catalysts. Various reviews give less than 1.0 m²/g [37, 54] to 0.03–0.06 m²/g [55]. A large, open macropore structure is also preferred. Although high-surface-area materials, such as molecular sieves, are mentioned in some patents, these are very questionable claims. The combination of low thermal conductivity inside large-surface-area catalysts coupled with the difficulty of product diffusion out of the pores is generally blamed for the poor performance of high-surface-area carriers. Supports are $\frac{1}{4}$- to $\frac{3}{8}$-in.-diameter spheres, cylinders, or rings. Pore volume is about 0.5 cm³/g, and median pore diameter is 20 μm.

5. Catalyst Preparation

There are two general methods of placing the silver on the carrier. One is the deposition of insoluble silver compounds on the outside of the support; this is frequently called deposition or coating. The other procedure uses soluble silver compounds and also saturates the inside of the support; it is called impregnation.

For deposition insoluble compounds of silver oxide, silver carbonate, and silver oxalate are used. These are precipitated from silver nitrate with the proper sodium salts of the anions and washed until they are anion- and cation-free. Other cations can also be used, such as Ba(OH)₂, but KOH use is questionable since washing has to be extremely thorough because even a very small amount of potassium can influence the performance. The washed insoluble silver salts in wet form (silver oxalate explodes when dry) are added to the carrier with the proper quantities of additives and heated during rotation until an even coating is obtained. Then drying and roasting follow. Deposited catalysts usually have a higher silver content and initial activity but are subject to a loss of selectivity during use, and dusting of silver tends to increase the pressure drop in converter tubes.

For impregnation the precipitated insoluble silver salts are dissolved in lactic acid [56] or complexed with ammonia [57] or amines [58, 59]. Detailed preparation examples are given in a recent U.K. patent application [48]. The carrier is soaked in the impregnating solution, sometimes with the application of a vacuum, and then excess solution is drained and the impregnated support is heated to decompose the silver complex. Heating time and temperature have an important effect. A short time at a high temperature is preferred.

D.   REACTION MECHANISM ON A PROMOTED CATALYST

1.   Action of Vapor Phase Inhibitors

The author operated a pilot plant in the early 1950s that was in an industrial complex. The results were erratic until one day a correlation was found between wind direction and performance. A similar observation had been made at Union Carbide Corporation in South Charleston, West Virginia, more than a decade earlier, as was learned later, where an active carbon bed for the air feed eliminated the erratic performance and stabilized it, but at a very poor selectivity level. This led to the conclusion that some contaminant in the air must be responsible for the increased selectivity, and a long search was started. More than 200 chemicals produced around South Charleston, West Virginia, were tested. More than half of them had a significant effect on the silver-catalyzed oxidation process, mostly bad. Chlorinated hydrocarbons were found to be outstanding in suppressing the total oxidation although slowing down the production somewhat. This observation led to the Union Carbide patent on the inhibitor application obtained by Law and Chitwood [50, 60].

Even in the late 1950s some of the old, and by now nonexistent, air-operated plants were influenced by wind direction in the South Charleston, West Virginia, area. When the wind blew from the direction of the vinyl chloride unit, the ethylene dichloride inhibitor feed had to be cut back to maintain production. Vinyl chloride was made by the cracking of ethylene dichloride. In contrast, when the wind blew from the polyethylene unit, which had to make frequent blow-downs, inhibitor feed had to be sharply increased to prevent temperature runaways. This was of course, before the days of environmental protection.

In all the above-mentioned cases of erratic performance caused by air pollutants, both conversion and efficiency (defined as the number of moles of ethylene oxide formed divided by the number of moles of ethylene consumed) varied up and down, usually in opposite directions. From this it follows that the effects are not permanent in nature, at least over a shorter time range. Yet they are not adsorption effects only but involve chemical reactions with solids, as will be shown next.

2. Inhibitor Reactions with the Catalyst

Law and Chitwood recognized in 1940 [60] that catalysts could be pretreated with compounds, like ethylene dichloride, that react with the silver surface from the vapor space and then treated with Ba, Ca, or Sr hydroxide solutions to form an active catalyst.

Margolis [38] reported the work of Stepanov et al. [61] on the sequence of processes occurring on catalysts containing chlorine or other metalloid impurities. Their suggested scheme was

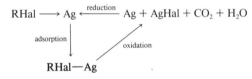

Although this scheme may be generally correct, it does not indicate what reduces AgHal to Ag in the oxidizing atmosphere. Endler [61] claimed that a catalyst, poisoned by an excess of inhibitor reaction, can be regenerated by adding methane or ethane to the reacting mixture. McNamee et al. [62] noticed that overchlorinated silver catalyst could be regenerated by treatment with ethylene oxide and water vapor. MacKim and Cambron [63] observed that "in the absence of paraffins the addition of traces of ethylene dichloride only served to poison the catalyst."

The previous five observations will be related on the basis of experiments made by the author almost 20 yr ago on a high-barium-containing deposited catalyst. Although this type of catalyst is obsolete today and has been replaced by better ones, the fundamental reactions are still the same even if their relative importance has changed.

First it was observed that on a catalyst operating at steady state for ethylene oxide production, a ppm quantity of vinyl chloride, equivalent to the ethylene dichloride consumed, was produced. Therefore the following reaction must occur:

$$Ag + C_2H_4Cl_2 + \tfrac{1}{2}O_2(Ag) \rightarrow C_2H_3Cl + AgCl + \tfrac{1}{2}H_2O$$

Most of the missing Cl was found in the product gas as ethyl chloride. In view of the oxidation conditions and the observations of Endler [62] this must be the result of the reaction

$$AgCl + \tfrac{1}{4}O_2(Ag) + C_2H_6 \rightarrow C_2H_5Cl + \tfrac{1}{2}H_2O + Ag$$

Endler states that CH_4 has the same effect, and McNamee *et al.* [63] claim the same effect for an ethylene oxide–water combination. These all go on, only the dechlorination rate with ethane is at least an order of magnitude higher than with the other compounds mentioned. Both vinyl chloride and ethyl chloride can be used as inhibitors, as well as any other chlorinated organic. The fate of the organic moiety cannot be observed with vinyl chloride, because in the reaction

$$Ag + C_2H_3Cl + \tfrac{1}{4}O_2(Ag) \rightarrow C_2H_2 + AgCl + \tfrac{1}{2}H_2O$$

the acetylene formed may be instantaneously oxidized. In a similar way the ethylene formed from ethyl chloride cannot be distinguished from the ethylene fed. Ethylene may have some dechlorinating effect itself, but it cannot be much, otherwise there would be no need for ethane. Some ethane is always present in the feed ethylene and, as Endler states, a few volume percent in the feed is all that is required. In oxygen-using technology large quantities of methane are added to the feed [64], and even ethane can be used [65]. The effect of the paraffinic hydrocarbon is more than that of the dechlorination—it increases the heat capacity, improves heat transfer, and thereby reduces the "hot spot" temperature. In addition it reduces the energy requirement for the cycle compressor, and it may involve the termination of chain reactions started on the catalyst surface but proceeding in the gaseous phase.

When ethyl chloride is used as an inhibitor over a catalyst operating at steady state with the air process where a low percentage of ethane is fed, virtually no change can be observed in the few parts per million of ethyl chloride concentration in feed and product gas, yet part of the ethyl chloride that leaves the reactor is not the same as that which was fed. This could be seen when deuterated ethyl chloride was used for the experiments, because

$$Ag + C_2D_5Cl + \tfrac{1}{4}O_2(Ag) \rightarrow C_2D_4 + AgCl + \tfrac{1}{2}D_2O$$

and

$$C_2H_6 + AgCl + \tfrac{1}{4}O_2(Ag) \rightarrow C_2H_5Cl + Ag + \tfrac{1}{2}H_2O$$

In actual experiments, where only deuterated ethyl chloride was fed, half of the discharged ethyl chloride had five hydrogens, proving that it came from the ethane.

On a high-barium-containing catalyst most of the chloride content of the catalyst is in the form of $BaCl_2$. A reaction analogous to the soda melt technology for recovering silver from silver chloride:

$$2AgCl + Na_2CO_3 \rightarrow 2NaCl + 2Ag + \tfrac{1}{2}O_2 + CO_2$$

occurs, in the solid phase of the catalyst, between silver and barium compounds and in a reversible manner:

$$2AgCl + BaCO_3 \rightleftharpoons BaCl_2 + 2Ag + \tfrac{1}{2}O_2 + CO_2$$

This reaction can be observed if, in an ethylene oxidation reaction that runs at steady state, where the amount of chlorides leaving the reactor is equal to the amount of chlorides fed, a sudden step change increase is made in the CO_2 concentration. After such a step change, for quite some time more chloride leaves the reactor than is fed. Also, when an operating system is suddenly flushed out with inert gas and the catalyst is leached out by water, a large quantity of soluble $BaCl_2$ is found and the pH of the solution is very high. This indicates that some of the chloride ions in the water extraction are readsorbed by the silver, otherwise no $Ba(OH)_2$ would be present since it would not survive the high CO_2 partial pressure of the reacting system. In the absence of Ag, $BaCl_2$ cannot be dechlorinated under ethylene oxidation reaction conditions. In the previous illustrations $\tfrac{1}{4}O_2(Ag)$ indicates only that oxygen chemisorbed on the silver surface participates in the reaction, without any regard as to stoichiometry or type of oxygen adsorbed. The role of adsorbed oxygen in the dechlorination of silver then explains why an overchlorinated surface is very hard to regenerate — there is practically no oxygen on the surface if the chlorine coverage is high. Meisenheimer and Wilson [66] have already observed that the rate of oxygen adsorption is 50 times slower when 25% of the surface is covered with chlorine than on a chlorine-free surface. Kilty et al. [52] observed the same and also that every additional chlorine atom over the 25% coverage prevented the adsorption of two more oxygen atoms. Therefore oxygen adsorption is impossible before all the sites are covered by chlorine.

In summary, the chlorine content of the catalyst is controlled and changed by the concentration of the chlorinated hydrocarbon, that of saturated hydrocarbons, and that of carbon dioxide in the reacting gas phase. In addition earth alkali metal compounds serve to store or buffer the chloride, exchanging chlorine with silver and making the silver less sensitive to overchlorination. Earth alkali metals make control of the conversion with the inhibitor easier. In all commercial processes the catalyst contains some alkali or earth alkali additive. In the reacting gases, besides the reactants and products of ethylene, oxygen, ethylene oxide, carbon dioxide, and water, hydrocarbon gases are always present, and inhibitors, usually chlorinated hydrocarbons, are always fed. Therefore any mechanistic or empirical kinetic model has to account for the effect of most of these components. The least important of these is water, followed by ethylene oxide and then perhaps ethylene itself, if it approaches the saturation level.

In the academic literature dealing with the reaction mechanism over pure silver without complications involving chlorinated hydrocarbons, the

Rideal–Eley mechanism, between adsorbed oxygen and gaseous ethylene, is favored over a Langmuir–Hinshelwood-type mechanism. Since, according to Ostrovskii et al. [51], the first fractions of chlorine increased the activity of the catalyst, the possibility of a reaction between ethylene adsorbed at the chlorinated sites with adsorbed oxygen cannot be excluded. Since AgCl is thermodynamically stable under reaction conditions, it is also possible that Cl ion migrates in the bulk of the silver, below the surface, leaving Ag^+ on the surface. Ethylene can be adsorbed at these sites and react with oxygen adsorbed on silver.

E. CATALYST TESTING

Silver catalysts produced for ethylene oxide manufacture are analyzed for their chemical composition, checked for their physical properties, and tested in a laboratory reactor for their catalytic activity.

The chemical composition is analyzed for silver, earth alkali metals, and alkali metals and for deleterious contaminants such as heavy metals, sulfur, and halogens. The physical property tests include BET measurement, usually done using Kr because of the small area. Porosity measurements for quality control can involve mercury porosimetry in spite of the known amalgamation tendency of silver because the oxygenated surfaces makes this process very slow. In tests for new research formulations, any and all of the modern high-vacuum surface techniques can be used. Most important are (ESCA), secondary ion mass spectroscopy (SIMS), and Auger methods.

For testing the activity in actual reactions, some companies still use small-diameter, thin-wall, coiled-tube reactors, at least for preliminary screening. In these empirical tests, there is no mathematically defined correlation between laboratory results and plant performance, therefore only companies who have a lot of practical experience, both commercially and in the laboratory, can make use of these methods [67]. For second-stage testing and for predicting plant performance, three major catalyst manufacturers in the United States use internal recycle reactors [68–70].

Some typical testing conditions using Berty reactors are described in a Union Carbide patent [48]. The conditions are catalyst volume charged, 80 cm³; feed rate, 22.6 SCFH = 178 cm³ (STP)/sec; space velocity, 8000/hr; pressure, 275 psig = 2.00 mP; rpm, 1500. The temperature is adjusted to receive 1% oxide in the produced gas (between 255 and 300°C). The feed composition is oxygen, 6.0 mol%; ethylene, 8.0 mol%; ethane, 0.5 mol%; carbon dioxide, 6.5 mol%; nitrogen, balance of the gas; ethylene dichloride, 7.5 ppm.

In this test, the outlet ethylene oxide concentration and reaction temperature are monitored for 4 to 6 days to make certain that the catalyst reaches its peak steady state performance. During this time, a 1% concentration of ethylene oxide is maintained by adjusting the temperature as required.

The final tests for performance evaluation are usually made in pilot plants that consist of one or a few tubes of the same size as those in the commercial unit. These tests are made with oxide removal in a closed cycle operation simulating the actual plant conditions as well as possible. These are expensive, slow tests and are limited to proving predictions from the recycle reactor [69, 70] and checking activity decline with use and, at times, catalyst life.

IV. Manufacturing Processes

A. PHYSICOCHEMICAL BASES OF ETHYLENE OXIDATION

Ethylene oxidation over silver proceeds in a very clean manner in the sense that no minor by-products are made; besides ethylene oxide the only by-products are carbon dioxide and water. The products and by-products are formed in the following three stoichiometric reactions, at 277°C, with the given heats and free energies of reactions:

$$C_2H_4 + \tfrac{1}{2}O_2 \rightarrow C_2H_4O$$

$$\Delta H° = -117 \quad kJ/mol, \qquad \Delta G° = -50.6 \quad kJ/mol$$

$$C_2H_4 + 3O_2 \rightarrow 2CO_2 + 2H_2O$$

$$\Delta H° = -1217 \quad kJ/mol, \qquad \Delta G° = -1249 \quad kJ/mol$$

$$C_2H_4O + 2\tfrac{1}{2}O_2 \rightarrow 2CO_2 + 2H_2O$$

$$\Delta H° = -1334 \quad kJ/mol, \qquad \Delta G° = -1294 \quad kJ/mol$$

According to the preceding, the chemical equilibrium for all three reactions heavily favors the products. Therefore, the reason ethylene oxide is not oxidized is purely kinetic. Also, it can be seen that the heat generated by the total oxidation reactions is an order of magnitude higher than the heat produced in the epoxidation.

The flammable limit of ethylene in air is about 3 vol%. As the three-component mixture becomes leaner in oxygen, the permissible ethylene content increases. For example, with 7% oxygen, about 9% ethylene is permissible. Also, with 30% ethylene, up to 9% oxygen can be tolerated because of the higher heat capacity. Considerations of how close one should approach the dangerous limit involve balancing this with the benefits in productivity and selectivity. The best results are achieved at higher concentrations, but a margin has to be left for errors in analysis and control. Since ethylene oxide concentration in the product gas is usually between 1 and 2 vol%, it does not increase the flammability problem. The previously mentioned limits are for the customary 1–2 MPa operating pressures, and the limits decrease with a further increase in pressure.

The reason for running the process under pressure is not based on equilibrium considerations, as can be concluded from free energy data presented previously. The main reason is energy conservation. The high reaction rate coupled with a high heat generation rate requires very good heat and mass transfer conditions. These can be achieved only at high mass velocities, and these in turn result in a high pressure drop over the converter tubes unless the process is operated at an increased pressure of 1 to 2 MPa. Recovery of the produced ethylene oxide is also more economical at a higher pressure.

B.   FEED MATERIALS

The purity requirements for both ethylene and air or oxygen feeds are quite high and have steadily increased in the past years. Ethylene purity is required to be above 97–98%. The usual ethylene contaminants, acetylene, CO, C_3 hydrocarbons, and hydrogen are harmful because they oxidize faster, overheating the catalyst and depositing cokelike material. Less than about 10 ppm of these compounds can be tolerated. Small quantities of ethane and methane are not harmful and can even be useful.

Early ethylene oxide plants operated with rather impure ethylene. With the growth of the polyethylene business, ethylene demand increased and polymerization technology required higher purity than the oxidation process. Yet it was only practical to have one quality of ethylene, especially in the merchant market served by ethylene pipelines. Ethylene oxide plants using better quality feeds realized significant improvements and then standardized on these purer feeds.

Air has to be free of volatile organic contaminants and of lubrication oil mists. Almost all oxygenated compounds and aromatics act as inhibitors. Chlorine and especially sulfur compounds are catalyst poisons. Although chlorine can be removed, sulfur poisoning is practically irreversible [50]. These contaminants should be kept below 1 ppm.

Oxygen is free of all contaminants because the safety of the air separation process requires it. For oxygen-based processes it should be above 98%

purity. The remainder is mostly argon and, although this element is inert, it accumulates in the recycle stream. The increase in the needed volume of the purge stream is proportional to the inert content, and, therefore, the amount of argon should be kept low too.

C. CATALYTIC OXIDATION TECHNOLOGY

1. The Catalytic Reactor

All present commercial processes operate with multitube, packed bed reactors. Patents exist that describe fluid bed reactors [71], transport line reactors [72, 73], and adiabatic reactors [74]. A detailed description of the German process developed during World War II that used a tubular reactor is available [16, 75, 76]. The reactor designed for the Zweckel plant is shown in Fig. 1.

Cooling of the reactors can be accomplished by cocurrently circulated, pressurized heat transfer fluids (Mobiltherm, Dowtherm, or tetralin) with a multiple-pass heat exchanger configuration using baffles. Cooling can also be achieved by boiling organic heat transfer fluids, and in this case no baffles are used in the shell. In both cases, the hot heat transfer fluid is cooled in a boiler, generating high-pressure steam. Newer, more active catalysts that operate at a lower temperature permit cooling with direct water boiling in the reactor shell [79]. The high-pressure steam generated by the reaction heat is usually enough to drive the recycle compressor by expanding to a lower pressure. The low-pressure steam in turn can drive all the pumps in the plant and then heat the boilers of the distillation towers.

The most important design consideration for ethylene oxidation reactors is the safe and economic removal of reaction heat while maintaining high productivity and good selectivity. Even with high mass velocities at 2 MPa of pressure, the limiting heat transfer coefficient is at the inner surface of the reactor tubes. The main task of the design is integration of the two differential equations for ethylene oxide and carbon dioxide formation rates simultaneously with equations for heat generation and mechanical energy consumption for a pressure drop.

The kinetic model is expressed in the form of rate equations that have to account for the effects of temperatures and partial pressures of the reactants, products, inhibitors, and paraffin hydrocarbons. The actual equations used for design are proprietary information of the companies and are not published. Equations published in the technical literature never account for the inhibitor and paraffin effects and are not directly useful in commercial reactor design.

For the integration procedure used in the reactor design, the simple,

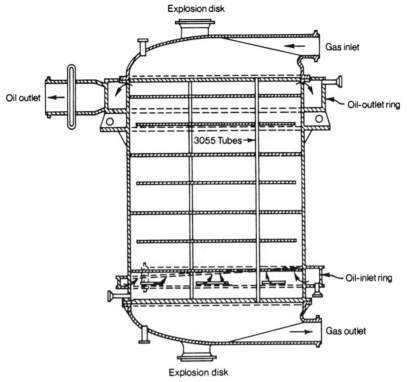

Explosion disk

Gas inlet

Oil outlet

Oil-outlet ring

3055 Tubes→

Oil-inlet ring

Gas outlet

Explosion disk

Fig. 1. Tubular reactor for ethylene oxidation. BIOS(1947).

one-dimensional, quasi-homogeneous approach is satisfactory. This is because the tube diameter/catalyst particle diameter ratio is small, as low as 3,
and therefore strong cross-mixing exists, eliminating most of the radial
concentration gradients and moderating the temperature peaks. The plug-
flow assumption is closely attained, since most of the temperature gradient
is at the inside wall of the reactor tube. Most important cases can be
checked by using a two-dimensional homogeneous model and integrating
partial differential equations in axial and radial directions by orthogonal
collocation or other methods. At the temperature peak, inside the tube, the
temperature should not be more than 30–40°C higher than the coolant
temperature for thermally stable operation [78]. This 30–40°C maximum
difference approximately satisfies the "slope condition" of the thermal
stability criteria [79, 80] expressed as

$$\Delta T_{max} = (RT^2/E)(C_0/C)$$

For the calculation of transient temperature changes the one-dimensional, two-phase model is used [81].

2. Air versus Oxygen

All early ethylene oxide plants were built for oxidation with air; the newest plants are almost entirely built for oxygen. Oxygen-based plants have been more economical than air-operated plants ever since their introduction in the early 1960s for smaller and medium-sized units. For the largest plants, air operation has remained more economical because energy recovery can be used more extensively.

For air-operated plants, the need for purge reactors is the greatest disadvantage. The need for high selectivity at high production rates requires the highest possible ethylene and oxygen concentrations and only their partial conversion. Therefore, the inert purge from these primary reactors contains a significant quantity of unconverted raw materials. This can be 10–20% of the feed, so it has to be effectively utilized for additional ethylene oxide production.

In order to avoid construction of a purge reactor different in size from the primary reactor, a large plant is needed that has 5–10 primary reactors running in parallel to feed one purge reactor. Utilization of the purge stream can actually involve a train of such reactors instead of one, the first few operating with their own recycle blowers and scrubber systems. The problem of air versus oxygen was discussed in detail by Gans [39].

Oxygen-operated plants need a special scrubbing system to remove CO_2 formed by total oxidation. Usually CO_2 removal is accomplished by a hot K_2CO_3 solution that can contain various additives. These additional investment and operating costs offset some of the advantages of the oxygen-based process. Because of a lower purge stream, oxygen-based plants can operate at higher ethylene concentrations and usually have a somewhat higher yield and productivity.

For oxygen plants it can be argued that about 70% of the oxygen cost is in air compression, which is also required for air-operated plants. This is only partially true; in oxygen production only a small part of the compression work can be recovered, whereas air-operated plants can "borrow" air, use up three-fourths of the oxygen content, and return the compressed air for energy recovery in a gas turbine. A simplified scheme is given in Fig. 2, and the possibilities for this are discussed by Canova [82] and Kydd [83] and in numerous patents such as those obtained by Parmegiani and Bellofatto [84].

In a scheme where gas turbines are used for energy recovery, both air compression and expansion can be carried out in two stages. The blow-off gas from the last purge stream scrubber is reheated for the first-stage

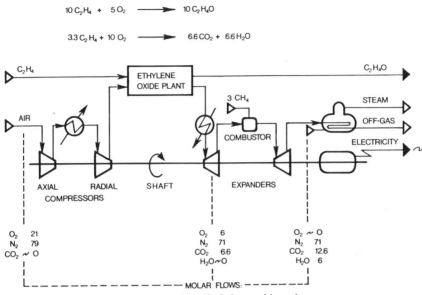

Fig. 2. Cogeneration of ethylene oxide and energy.

expansion in a heat exchanger by hot process gas, and for the second-stage expansion in a combustor. This combustion then takes care of pollution control problems for the process gas, and this combustion can also be accomplished catalytically.

The borrowing of air is made possible by the temperature limitation in the hot gas expander phase of the Brayton cycle. Present-day gas turbine expanders are limited to 600–650°C maximum temperature. To reach this level, only about 6% oxygen from the compressed air is used up. The rest of the oxygen is just working fluid, like the nitrogen and combustion gases, and therefore it can be extracted by the process. Some of it is replaced by the by-product CO_2, so over 85% of the compressed air, with only about a 0.3- to 0.4-MPa pressure loss, is available for energy production, and only the volume and pressure losses are debits in the ethylene oxide process. Systems built optimally for the coproduction of ethylene oxide and electrical energy have capital requirements in addition to those of ordinary air-operated plants, but they would have done well economically, since in the 1970s energy prices increased relatively more than ethylene oxide prices. However, large-tonnage oxygen-supplied plants can compete even with these operations. More depends on nitrogen and argon by-product utilization and on the capitalization of oxygen plants than on the technology differences between air and oxygen operations.

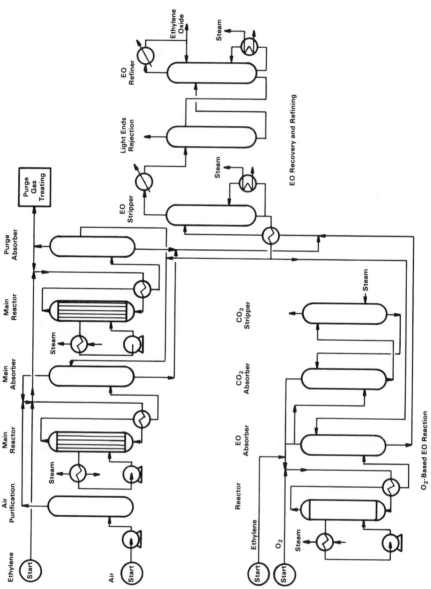

Fig. 3. Ethylene oxide (EO) technology of Scientific Design Company. Reprinted by permission of Scientific Design Company, a division of the Halcon SD Group, Inc., New York.

Flowsheets for air and oxygen plants are shown in Fig. 3. In both cases, the ethylene oxide is washed out by scrubbing with water. Ethylene oxide is stripped out of the lean water solution, concentrated, and refined. Acetaldehyde and some condensation products are the main impurities to be removed. Some of these are produced in overheated tubes, and others are formed in liquid phase reactions during separation.

Increasing impurities in the product point toward overheated conditions in some of the tubes. In overheated tubes, a runaway condition exists. Oxygen is the limiting reactant, and it is consumed at the hot spot. Some ethylene oxide survives these conditions, only to be isomerized under reducing conditions on the oxygen-free silver surface downstream from the hot spot. From a few "hot" tubes acetaldehyde can accumulate to the level where it can start a chain oxidation reaction in the empty space of the discharge head, causing a sudden increase in temperature. These degenerate explosions, as they are called in free radical kinetics, do not cause damage or even rupture disk failure but severely upset the steady rate of production.

D.   FUTURE OUTLOOK

Several new technologies have emerged in the past 20 yr to replace the silver-catalyzed oxidation of ethylene to ethylene oxide. Because glycol production is the single largest use for ethylene oxide, some of these processes should be reviewed also.

Union Carbide [85] patented an arsenic-catalyzed liquid phase process for olefin oxides. The process has very high selectivity at 60–130°C, and no degradation of the 1,4-dioxane solvent was observed. The high cost of hydrogen peroxide makes this approach presently uneconomical.

Halcon has patented [86, 87] a thallium-catalyzed process for making epoxides. The reoxidation process with oxygen is very slow. Faster reoxidation can be achieved by organic peroxides, which again makes it expensive.

Oxirane built a large plant for producing glycol from ethylene via the acetoxidation process using a tellurium catalyst [88]. Although the yields were good, the plant was shut down after less than 2 yr of operation because of severe problems involving corrosion and keeping the volatile catalyst component confined to the reactors.

Union Carbide is developing a direct route from synthesis gas to glycol using a rhodium carbonyl complex catalyst [89]. The necessary pressures for the reaction, in the 100–200 MPa range, are presently too high to make this process feasible.

Texaco Chemical made significant improvements in ruthenium-catalyzed glycol synthesis [90] that may lead to commercialization of the synthetic gas route. By using ruthenium melt catalyst the synthesis pressure could be kept at the 20- to 40-MPa level.

References

1. R. E. Parker and N. S. Isaacs, *Chem. Rev.* **59,** 737 (1959).
2. H. Fujimoto *et al., Bull. Chem. Soc. Jpn.* **49,** 1508 (1976).
3. C. A. Coulson, "Valence." Oxford Univ. Press, New York (1976).
4. C. L. Yaws and M. P. Rackley, *Chem. Eng.* , 129 (1976).
5. L'Air Liquide "Gas Encyclopaedia." Elsevier, Amsterdam (1976).
6. Union Carbide Corp. Brochure F-41515B, Glycols (1978).
7. J. Farakawa and T. Saequsa, "Polymerization of Aldehydes with Oxides." Wiley (Interscience), New York (1963).
8. T. H. Baize, *Ind. Eng. Chem.* **53,** 93 (1961).
9. Kirk-Othmer, "Encyclopedia of Chemical Technology," 2nd ed., Vol. 8, p. 545. (Interscience), New York (1965).
10. Kirk-Othmer, "Encyclopedia of Chemical Technology," 3rd ed., Vol. 9, p. 440. Wiley (Interscience), New York (1980).
11. Chemical Marketing Reporter July 7 (1980).
12. Chemical Marketing Reporter Sept. 14 (1981).
13. A. Wurtz, *Ann.* **110,** 125 (1859); *Ann. Chim. Phys.* **55,** 433 (1859).
14. A. Wurtz, *Ann. Chim. Phys.* **69,** 355 (1863).
15. T. E. Lefort, Fr. Pat. 729,952 (1931); U.S. Pat. 1,998,878 (1935).
16. FIAT Field Intelligence Agency Technical Final Rep. 875 (1947).
17. P. Schutzenberger, *Bull. Soc. Chim.* **31,** (2) 482 (1879).
18. R. Willstatter and M. Boomer, *Ann.* **422,** 36 (1921).
19. R. C. Lemon, P. C. Johnson, and J. M. Berty, U.S. Pat. 3,132,156 (1964).
20. A. W. Czanderna, *J. Phys. Chem.* **70,** 2125 (1966).
21. S. Dushman, "Scientific Foundations of Vacuum Technique," 2nd ed., p. 749. Wiley, New York (1962).
22. V. E. Ostrovskii and M. I. Temkin, *Kinet. Catal.* **7,** 466 (1966).
23. A. W. Czanderna, *J. Phys. Chem.* **68,** 2765 (1964).
24. A. W. Czanderna, *Thermochim. Acta* **24,** 359–367 (1978).
25. W. M. H. Sachtler, *Catal. Rev.* **4,** 27 (1970).
26. M. N. P. Janssen, J. Moolhuysen and W. M. N. Sachtler, *Surf. Sci.* **33,** 625 (1972).
27. X. E. Verykios, F. P. Stein, and R. W. Coughlin, *Catal. Rev. Sci. Eng.* **22**(2), 197 (1980).
28. A. E. Marcinkowsky and J. M. Berty, *J. Catal.* **29,** 494 (1973).
29. E. L. Force and A. T. Bell, *J. Catal.* **38,** 440 (1975).
30. E. L. Force and A. T. Bell, *J. Catal.* **40,** 356 (1975).
31. A. F. Benton and J. C. Elgin, *J. Am. Chem. Soc.* **51,** 7 (1929).
32. G. H. Twigg, *Proc. R. Soc. London* **A188,** 92, 105, 123 (1946).
33. H. Worbs, U.S. Office of the Publication Board, Rep. 98,705 (1942).
34. M. T. Meter, U.S. Office of the Publication Board, Rep. 73530 (1945).
35. G. R. Schultze and H. Thiele, *Erdoel Kohle* **5,** 552 (1952).
36. S. Carra and P. Forzatti, *Catal. Rev. Sci. Eng.* **15**(1), 1–52 (1977).
37. H. H. Voge and C. R. Adams, *Adv. Catal.* **17,** 151 (1967).
38. L. Ya. Margolis, *Adv. Catal.* **14,** 429 (1963).

39. M. Gans, *Chem. Eng. Prog.* **75**(1), 67 (1979).
40. E. McKeown and W. A. Waters, *J. Chem. Soc. (London)*, **B1040** (1966).
41. W. M. H., Sachtler, C. Backx, and R. A. Vansanten, *Catal. Rev. Sci. Eng.* **23**(1–2) 127–129 (1981).
42. M. I. Temkin, *Adv. Catal.* **28**, 173 (1979).
43. A. W. Czanderna, *J. Colloid Interface Sci.* **24**, 500 (1967).
44. I. E. Maxwell, U.S. Pat. 4,033,093 (1977).
45. G. H. Law, U.S. Pat. 2,142,948 (1939).
46. J. Bathory, A. Balogh, and I. Hartwig, *Proc. Conf. Appl. Phys. Chem.* **2**, 279 (1971).
47. J. J. Carberry, G. C. Kuczynski, and A. Martinez, *J. Catal.* **26**, 247 (1972).
48. M. M. Bhasin, P. C. Ellgen, and C. H. Hendrix, U.K. Pat. Appl. 2,043,481 (1980).
49. H. T. Spath, G. S. Tomazic, H. Wurm, and K. Tokar, *J. Catal.* **26**, 18 (1972).
50. G. H. Law and H. C. Chitwood, U.S. Pat. 2,279,469 (1942).
51. N. V. Ostrovskii, N. V. Kul'kova, V. L. Lopatin, and M. I. Temkin, *Kinet. Catal.* **3**, 160 (1962).
52. P. A. Kilty, N. C. Rol, and W. M. H. Sachtler, *Proc. Int. Congr. Catal. 5th.* **2**, 929 (1973).
53. G. W. Sears, U.S. Pat. 2,605,239 (1952).
54. J. K. Dixon and J. E. Longfield, *Catalysis* **7**, 248 (1960).
55. J. Bathory and A. Balogh, *Magyar Kemikusok Lapja* (Hungarian Journal of Chemists) **XVIII**, 12 (1962).
56. R. S. Aries, U.S. Pat. 2,477,435 (1949).
57. T. J. West and J. P. West, U.S. Pat. 2,463,229 (1949).
58. G. Schwartz, U.S. Pat. 2,459,896 (1949).
59. R. P. Nielsen, U.S. Pat. 3,702,259 (1972).
60. G. H. Law and H. C. Chitwood, U.S. Pat. 2,194,602 (1940).
61. Yu. N. Stepanov, L. Ya. Margolis, and S. Z. Roginskii, *Kinet. Katl.* **2**, 684 (1960).
62. H. Endler, Italian Pat. 600,394 (1959).
63. F. L. W. McKim and A. Cambron, *Can. J. Res.* **B27**(11), 813 (1949).
63. R. W. McNamee, H. C. Chitwood, and G. H. Law, U.S. Pat. 2,219,575 (1940).
64. H. A. Kingsley and F. A. Cleland, U.S. Pat. 3,110,837 (1964).
65. D. Brown, Belg. Pat. 707,567 (1968).
66. R. G. Meisenheimer and J. N. Wilson, *J. Catal.* **1**, 151 (1962).
67. J. V. Porcelli, *Catal. Rev. Sci. Eng.* **23**(1–2), 151–162 (1981).
68. J. M. Berty, *Chem. Eng. Prog.* **70**(5), 78–83 (1974).
69. J. M. Berty, *Catal. Rev. Sci. Eng.* **20**(1), 75–96 (1979).
70. J. M. Berty, *Chem. Eng. Prog.* **75**(9), 48–51 (1979).
71. T. E. Corrigan, *Pet. Refiner.* **32**(2), 87–89 (1953).
72. J. M. Berty, Austrian Pat. 201,575 (1959).
73. J. M. Berty, German Pat. 1,068,684 (1960).
74. J. M. Berty, German Pat. Application 1,915,560 (1969).
75. CIOS Report No. 27/85. Combined Intelligence Objectives Subcommittee (1947).
76. BIOS Report No. 360. British Intelligence Objectives Subcommittee (1947).
77. J. C. Zomerdijk and M. W. Hall, *Catal. Rev. Sci. Eng.* **23**(1–2) 163–185 (1981).
78. G. K. Boreskov, R. N. Vasilevich, R. N. Gur'yanova, and B. B. Chesnokov, *Kinet. Catal.* **3**, 182 (1962).
79. D. D. Perlmutter, "Stability of Chemical Reactors." Prentice–Hall, Englewood Cliffs, New Jersey (1972).
80. J. M. Berty, J. P. Lenczyk, and S. M. Shah, *AIChE J.* to be published (1983).
81. J. M. Berty, J. H. Bricker, S. W. Clark, R. D. Dean, and T. J. McGovern, *Chem. Reac. Eng. Proc. Eur. Sym. 5th* **B–8** (1972).

82. F. Canova, *Chem. Eng.* 179–182 (1969).
83. P. H. Kydd, *Chem. Eng. Prog.* **71**(10), 62–68 (1975).
84. M. Parmegiani and D. Bellofatto, U.S. Pat. 3,552,122 (1971).
85. C. H. McMullen, U.S. Pat. 3,993,673. (1976).
86. W. F. Brill, U.S. Pat. 4,021,453 (1977).
87. N. Rizkalla, U.S. Pat. 4,058,542 (1977).
88. S. C. Johnson, *Chem. Eng. Prog.* **72**(9), 25 (1976).
89. R. L. Pruett and W. E. Walker, U.S. Pat. 3,833,634 (1974).
90. J. Knifton, *J. Am. Chem. Soc.* **103**(13), 3959 (1981).
91. C. J. Walters and J. M. Smith, *Chem. Eng. Prog.* **48**, 337 (1952).
92. A. S. Pell and G. Pichler, *Trans. Faraday Soc.* **61**, 71 (1965).
93. W. F. Giauque and J. Gordon, *J. Am. Chem. Soc.* **71**, 2176 (1949).
94. R. W. Gallant, *Hydrocarbon Process.* **46**(3), 143 (1967).
95. C. L. Yaws, "Physical Properties," McGraw–Hill, New York (1977).
96. O. Mass and E. H. Boomer, *J. Am. Chem. Soc.* **44**, 1709 (1922).
97. D. R. Stull, E. F. Westrum, and G. C. Sinke, "The Chemical Thermodynamics of Organic Compounds," p. 419. Wiley, New York (1969).
98. K. A. Kobe and R. E. Pennington, *Pet. Refiner* **29**(9) 135 (1950).
99. Union Carbide Corp. Brochure F-7618E, Ethylene Oxide (1973).
100. W. W. Smeltzer, E. L. Tollefon, and A. Cambron *Can. J. Chem.* **34**, 1046 (1956).
101. Y. L. Sandler, S. Z. Beer, and D. D. Durigon *J. Phys. Chem.* **69**, 4201.
102. Y. L. Sandler, S. Z. Beer, and D. D. Durigon, *J. Phys. Chem.* **70**, 3881 (1966).
103. W. Kollen and A. W. Czanderna, *J. Colloid Interface Sci.* **38**, 152 (1972).
104. G. Rovida, F. Ferroni, M. Maglietta, and F. Pratesi, *Surf. Sci.* **43**, 230 (1974).
105. A. W. Czanderna, S. C. Chen, and J. R. Beiger, *J. Catal.* **33**, 163 (1974).
106. A. F. Benton and L. C. Drake, *J. Am. Chem. Soc.* **56**, 255 (1934).
107. C. B. McCarty, Ph.D. Thesis, Purdue Univ. Univ. Microfilms 61–5741.
108. N. V. Kulkova and M. I. Temkin, *Russ. J. Phys. Chem.* **36**, 931 (1962).
109. R. B. Clarkson and A. Cirillo, *J. Catal.* **33**, 392 (1974).
110. A. I. Kurilenko, N. V. Kulcova, L. P. Baranova, and M. I. Temkin, *Kinet. Catal.* **3**, 177 (1962).
111. K. E. Murray, *Aust. J. Sci. Res. Ser.* **A3**, 433 (1950).
112. S. Wan *Ind. Eng. Chem.* **45**, 234 (1953).
113. R. E. Kenson and M. Lapkin *J. Phys. Chem.* **74**, 1493 (1970).
114. G. Liberti, A. Mattera, F. Pedretti, N. Pernicone, and S. Sattini, *Atti Accad. Naz. Lincei Cl. Sci. Fis. Mat. Nat. Rend.* **S2**(3) 392 (1972).
115. V. N. Korchak and I. I. Tret'yakov, *Kinet. Catal.* **18**(1), 141 (1977).

CHAPTER 9

Oxychlorination of Ethylene

J. S. Naworski   E. S. Velez

Richmond Research Center
Stauffer Chemical Company

I. Introduction

This chapter deals with the catalytic oxychlorination of ethylene to form ethylene dichloride (EDC). In vinyl chloride monomer (VCM) plants, EDC is dehydrohalogenated (cracked) to form VCM which is the monomer used in the production of polyvinyl chloride (PVC). Ethylene oxychlorination processes were developed commercially in the mid-1960s and provided an efficient mechanism for consuming the large quantities of HCl formed in the cracking sections of VCM plants. In this chapter, various ethylene oxychlorination processes are summarized. Some major fluidized bed processes are reviewed, but fixed bed oxychlorination is emphasized.

A. IMPORTANCE OF THE VCM-PVC INDUSTRY

Virtually all the VCM produced worldwide is used in the manufacture of PVC or vinyl copolymers. Thus, the historical importance of VCM manufacture parallels the phenomenal growth of the PVC industry. For 1981, worldwide VCM capacity was estimated at 42 billion lb/yr compared to about 6 billion lb/yr in 1964. The growth rate over this 17-yr period has averaged 12%, although the rate of growth has slowed in recent years. Most VCM continues to be produced in Western Europe, the United States, and Japan, but significant new capacity is being installed in industrially emerging countries and in the OPEC nations. Continued, but modest, PVC growth rates are predicted for the 1980s as per capita PVC consumption rises in underdeveloped countries and new uses are found elsewhere.

B. HISTORICAL DEVELOPMENT OF ETHYLENE OXYCHLORINATION PROCESSES

Before the development of ethylene oxychlorination processes, VCM was prepared by acetylene hydrochlorination or by direct chlorination of ethylene followed by cracking of the EDC. In the acetylene process, hydrochloric acid produced by combusting hydrogen with chlorine was reacted with carbide-derived acetylene [1]. This vapor phase, fixed bed reaction was usually conducted on a mercuric chloride catalyst deposited on activated carbon.

$$CH\equiv CH + HCl \xrightarrow{HgCl_2} CH_2=CHCl$$

Raw materials for this reaction were expensive even after petroleum-derived acetylene and by-product HCl came into common use.

One processing scheme that reduced the high raw material costs associated with acetylene hydrochlorination was the balanced ethylene–acetylene VCM process [1]. In the first step of this process, EDC is produced by the direct chlorination of ethylene

$$C_2H_4 + Cl_2 \rightarrow CH_2ClCH_2Cl$$

The EDC is then thermally cracked to form VCM and HCl.

$$CH_2ClCH_2Cl \xrightarrow{500°C} CH_2=CHCl + HCl$$

Finally, the HCl produced in the second reaction reacts with acetylene to produce additional VCM.

$$CH\equiv CH + HCl \xrightarrow{HgCl_2} CH_2=CHCl$$

Approximately one-half of the VCM is derived from acetylene and half from the less expensive ethylene. The HCl produced in the dehydrochlorination of ethylene dichloride is completely consumed to produce more VCM. Thus there is no problem with disposing of HCl, and raw material costs are lower than with the acetylene-based process.

The demise of acetylene-based VCM manufacture was signaled when balanced VCM processes based on ethylene oxychlorination were developed. The chemistry is described subsequently. The first two steps are similar to the ethylene–acetylene process, i.e., direct chlorination of ethylene and dehydrochlorination of the resulting EDC:

$$C_2H_4 + Cl_2 \rightarrow CH_2ClCH_2Cl$$

$$CH_2ClCH_2Cl \rightarrow CH_2{=}CHCl + HCl$$

The key step is the oxychlorination (or oxyhydrochlorination) of ethylene using the HCl produced in the cracking step.

$$C_2H_4 + \tfrac{1}{2}O_2 + 2HCl \xrightarrow{\text{CuCl}_2} CH_2ClCH_2Cl + H_2O$$

This vapor phase reaction can be conducted on a copper chloride catalyst in either a fixed or fluidized bed. The commercialization of ethylene oxychlorination revolutionized the VCM (and PVC) industry in the mid-1960s. Vinyl chloride could be made from inexpensive, petroleum-derived ethylene and chlorine. The process could be balanced with regard to HCl; all the HCl produced in the cracking step was consumed in the oxychlorination step.

In April, 1964, B.F. Goodrich commercialized their fluidized bed ethylene oxychlorination process in a plant rated at 300 million lb/yr of EDC. Gaseous ethylene, HCl, and air were reacted in a fluidized bed of copper chloride-impregnated alumina particles. The process was successfully integrated into an existing VCM plant, making the overall process a balanced operation. In recognition of this development, B.F. Goodrich was awarded honorable mention in the 1965 Kirkpatrick Award competition [2].

Several other companies commercialized ethylene oxychlorination technology at about the same time as Goodrich. Stauffer Chemical Company developed an air-based, fixed bed system consisting of three tubular reactors in series. The catalyst was a high-surface-area alumina impregnated with copper chloride and potassium chloride. The process was commercialized in 1965 by the American Chemical Company (a joint venture of Stauffer and Atlantic-Richfield) in Long Beach, California. PPG Industries, another oxychlorination pioneer, developed a fluidized bed process that used oxygen instead of air. PPG fluidized the catalyst in multiple tubes, whereas Goodrich's fluid bed was in a single vessel. Other early ethylene oxychlorination processes were developed by Dow, Toyo Soda, Monsato, Rhone-Poulenc, Ethyl, and Frontier Chemical (now Vulcan). Most of these companies have since become major VCM producers and/or licensors of VCM technology.

C. BALANCED VCM PROCESS

Figure 1 is a simplified process flow diagram for a balanced VCM process with ethylene oxychlorination. The three major component processes — direct chlorination of ethylene, ethylene oxychlorination, and dehydrochlorination or cracking of EDC — are highlighted. Each of the component processes has undergone extensive development since balanced VCM plants were first commercialized. Consequently, several variations of each of the component processes are commercially practiced. The most representative of these variations are described in the following text.

1. Direct Chlorination of Ethylene

The first step in the balanced VCM process is the direct chlorination of ethylene to form EDC:

$$C_2H_4 + Cl_2 \rightarrow C_2H_4Cl_2$$

There are two fundamental variations of this process: low-temperature chlorination (LTC) and high-temperature chlorination (HTC). Both these

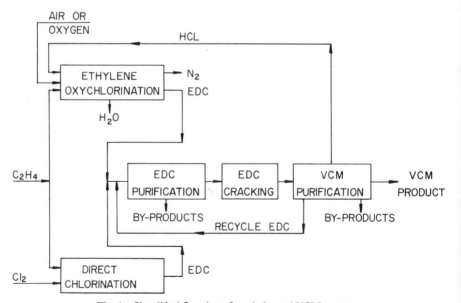

Fig. 1. Simplified flowsheet for a balanced VCM process.

processes have several variations because the technology has been evolving for many years.

In LTC, the chlorination of ethylene is conducted in a liquid phase of ethylene dichloride containing low concentrations of ferric chloride as catalyst. The reactor is equipped with an externally cooled heat exchanger that maintains reaction temperatures below the boiling point of EDC. The chlorine/ethylene ratio is approximately stoichiometric, with most operators favoring a slight excess of chlorine. Small amounts of oxygen or air are added to suppress by-product formation, particularly trichloroethane. The ferric chloride concentration is usually below 100 ppm. Chlorine is quantitatively converted, with selectivities to EDC of up to 99%. Because of the mild operating conditions, most of the equipment is carbon steel and operating problems are minimal. The LTC-produced EDC is distilled in a separate section of the balanced VCM process.

The distinguishing feature of HTC operation is that the heat of the chlorination reaction may be used to vaporize and distill EDC from the reaction liquid. In essence, the chlorination reactor is the reboiler for the distillation column. Because the heat of reaction is about six times that of the heat of EDC vaporization, EDC from oxychlorination and unconverted EDC from the cracking furnaces can also be distilled. Thus, HTC substantially reduces the total energy requirements for VCM production. Reactor types range from baffled or open tanks (Union Carbide) to circulatory, gas lift reactors (Stauffer).

High-temperature chlorination temperatures vary from the atmospheric boiling point of EDC (83°C) to about 130°C. The liquid medium consists primarily of EDC, with other chlorinated C_2 hydrocarbons as the predominant impurities. Some of the reactant liquid is periodically removed to maintain the desired liquid composition. As with LTC, the reaction generally is catalyzed by ferric chloride. Selectivities are somewhat insensitive to the ferric chloride concentration. Gaseous or liquid chlorine may be used, and slight ethylene excesses are typically employed. Oxygen (or air) is used to suppress side reactions. Selectivities of chlorine to EDC are 99+%, in Stauffer's HTC technology. By-product formation need not be higher than with LTC operation, even though the reaction temperature is significantly higher. Ethylene dichloride product purities are 99+% and are governed by the design and operation of the rectification section.

2. Dehydrochlorination or Cracking of EDC

In the cracking section of a VCM plant, VCM is produced from EDC in a vapor phase reaction in tubular pyrolysis furnaces. Large quantities of heat are required to split the HCl from the EDC molecule.

$$C_2H_4Cl_2 \xrightarrow[\text{heat}]{500°C} HCl + C_2H_3Cl$$

Although both catalytic and noncatalytic EDC pyrolysis are practiced, noncatalytic thermal cracking dominates commercially. Furnace design and operating parameters vary from plant to plant, but a number of characteristics are common.

Most plants operate at from 50 to 60% EDC conversion per pass. To achieve this conversion, the temperature of the gases exiting the furnace tubes is controlled at approximately 500°C. The axial temperature profiles along the tube length depend on furnace design and operating philosophy. Some plants heat the inlet gases quickly and maintain a roughly uniform temperature along the tube length; others heat the gases more slowly. Because the cracking reaction is highly endothermic, obtaining the desired temperature profile is difficult—particularly if there are other process upsets. Consequently, burner design, placement of the burners in the furnace, and firing patterns vary considerably from plant to plant.

High cracking temperatures increase the cracking depth at the expense of increased by-product and carbon formation. Process downtime for cleaning the furnace tubes is more frequent. EDC must be purified (distilled) and dried before being fed to the cracking furnaces. Most of the EDC that is not converted in the furnaces is reprocessed and recycled back to the cracking furnace. Thus, there is an economic trade-off among VCM yield, purity, downtime, production rates, and EDC reprocessing costs.

Ethylene dichloride cracking catalysts have the promise of reducing operating temperatures without decreasing conversions. Conversely, a catalyst could be used to increase the conversion at a given temperature. In early VCM plants, pumice and charcoal were used as cracking catalysts [3]. Wacker-Chemie reportedly has used a cracking catalyst of barium chloride on activated carbon [4]. Many patents describe cracking catalysts and promoters, but it is difficult to determine which ones have been used commercially. Catalysts and promoters most frequently mentioned are graphite, activated carbon, metallic chlorides (e.g., $CuCl_2$, $ZnCl_2$), chlorine, carbon tetrachloride, bromine, iodine, and various halogenated alkanes.

3. Ethylene Oxychlorination

Ethylene oxychlorination is the heart of modern-day balanced VCM processes. Oxychlorination balances the VCM process by consuming the HCl produced in the cracking section to produce more EDC. Overall, the raw materials for VCM production are ethylene and chlorine; no major quantities of by-products (other than water) are formed.

Although many ethylene oxychlorination processes have been developed, they can be distinguished by two main characteristics: whether air or pure oxygen is the source of oxygen and whether fixed or fluidized bed reactors are used. Table I lists these distinguishing features for some of the more important commercial oxychlorination processes. All the processes use a

TABLE I

Summary of Important Ethylene Oxychlorination Processes

Company	Source of oxygen	Reactor bed
Dow Chemical	Air	Fixed
Ethyl Corporation	Air	Fluidized
B.F. Goodrich	Air	Fluidized
Mitsui Toatsu	Oxygen	Fluidized
Monsanto	Air or oxygen	Fluidized
PPG	Oxygen	Fluidized
Rhone-Poulenc	Air	Fluidized
Stauffer	Air or Oxygen	Fixed
Toyo Soda	Air	Fixed
Vulcan	Air	Fixed

Deacon catalyst, with $CuCl_2$ being the primary active constituent. Several fluidized bed processes are described briefly in Section III. Section IV gives an in-depth presentation of fixed bed oxychlorination technology, with an emphasis on the catalyst and control of the reaction temperature.

II. Process Chemistry — Ethylene Oxychlorination

In commercial ethylene oxychlorination reactors, gaseous ethylene, HCl, and air (oxygen) catalytically react at temperatures in excess of 200°C to produce EDC. The overall chemical equation is

$$C_2H_4 + 2HCl + \tfrac{1}{2}O_2 \rightarrow C_2H_4Cl_2 + H_2O$$

Typically, the catalyst is composed of cupric chloride supported on high-surface-area alumina. Although other supports such as graphite, silica gel, calcined Fuller's earth, diatomaceous earth, pumice, or kieselguhr may be used, alumina is generally preferred because of its performance, attrition resistance and the ability to control its surface area [5]. Other metal salts, such as potassium, sodium, or aluminum chloride, may be added to the catalyst to increase selectivity and reduce volatilization of the copper chloride [6]. Fluid bed catalysts are made from alumina powder or microspheres which are about $20-100 \ \mu m$ in diameter. Fixed bed catalysts are usually tableted cylindrical pellets, extrudates, or spheres that are about $\tfrac{1}{8}-\tfrac{1}{4}$ in. in diameter.

The role of the copper chloride catalyst and the mechanism of the ethylene oxychlorination reaction has evolved. There is general agreement that the oxychlorination reaction does not proceed by the classic Deacon reaction, i.e., oxidation of HCl, followed by chlorination of ethylene.

In an early study of the reaction mechanism, Allen [7] proposed the following reactions:

$$CuCl_2 + \tfrac{1}{2}O_2 \rightarrow CuO + Cl_2$$
$$CuO + 2HCl \rightarrow CuCl_2 + H_2O$$

This sequence was based on Allen's thermodynamic calculations.

A kinetics study by Carrubba [8] tested possible kinetic expressions against his rate data at 184°C. An excellent fit was obtained with a rate expression based on a surface reaction between ethylene and oxygen. Although not completely satisfied, Carrubba proposed a reaction sequence involving reaction of ethylene with oxygen.

Arganbright [9] showed that olefins could be directly chlorinated with a fluidized bed of supported cupric chloride.

$$\diagup\!\!\!\!\diagdown C = C \diagdown\!\!\!\!\diagup + 2CuCl_2 \longrightarrow Cl-C-C-Cl + Cu_2Cl_2$$

HCl and oxygen are required to reoxidize the copper chloride.

$$Cu_2Cl_2 + 2HCl + \tfrac{1}{2}O_2 \rightarrow 2CuCl_2 + H_2O$$

Spector *et al.* [10] studied the ethylene oxychlorination reaction in an aqueous medium. The reaction proceeded very well when some cuprous chloride was added to complex the ethylene and brought into solution. As a follow-up to this work, Heinemann [27] showed that the induction time in the heterogeneous reaction could be significantly reduced by adding cuprous chloride to the cupric chloride catalyst. This work implies that a complexed ethylene – cuprous chloride intermediate participates in the reaction.

In commercial installations, HCl conversions are essentially complete and EDC selectivities in both fixed and fluid beds are high. Demonstrated EDC yields exceed 96% from ethylene and 98% from HCl. The most common chlorinated by-products are 1,1,2-trichloroethane, carbon tetrachloride, chloroform, chloral, and ethyl chloride. Most of the materials are separated from EDC in the various purification steps before cracking. Depending on local conditions, some plants sell ethyl chloride and/or 1,1,2-trichloroethane.

Because the oxychlorination reaction is highly exothermic, substantial quantities of heat must be removed from the reactors. In fluidized beds, the heat is removed by internal cooling coils that are submerged in the fluid bed [11]. Reaction temperatures are generally controlled in the range 210–240°C. Because the fluid bed is essentially isothermal, reaction conditions are uniform throughout the bed. An optimum reactor temperature can be achieved by proper design and operation of the heat removal system. For a given system, maintenance of the reaction temperature is thus not highly dependent on the activity of the catalyst. Catalyst selection (or development) is based primarily on attrition resistance, fluidization properties, and selectivity. Fluidized bed catalysts must have good structural integrity so that excessive amounts of fines are not generated. Care must be taken to avoid sticky catalyst particles because catalyst agglomeration can lead to poor fluidization or loss of fluidization entirely.

The catalyst temperature in fixed bed reactors varies continuously along the length of the reactor. The reaction tubes, which are about 1-in. in diameter, are surrounded by a cooling medium, usually boiling water, in a shell-and-tube configuration. As the preheated gases enter a reaction tube, heat is generated in proportion to the rate of reaction. Heat is removed by the coolant, which is approximately at a constant temperature along the entire length of the reaction tube. If the catalyst at the entrance of a reactor is too active, heat will be generated faster than it is removed through the reactor walls. The temperature of the packed bed will increase until the rate of heat removal through the walls equals the rate of heat generation. The axial temperature profile will have a maximum temperature or "hot spot" at some location in the reactor tube. Properly controlled reaction temperatures lead to good selectivities, long catalyst life, and a low pressure drop across the reactor.

Control of reaction temperatures is the most important concern in operating fixed bed reactors. Because the reaction temperature is directly related to catalyst activity, operation of a fixed bed reactor is more dependent on catalyst activity than operation of a fluid bed system. Moderation of the hot spots is achieved by reducing the reaction rate per unit volume of the reactor. Two successful approaches are to either dilute the catalyst by mixing it with inert particles or to vary the salt concentration of the catalyst in different sections of the reactor. These concepts will be discussed in more detail in Section IV.

III.   General Ethylene Oxychlorination Process Descriptions — Fluidized Beds

A.   B.F. GOODRICH FLUIDIZED BED OXYCHLORINATION

The B.F. Goodrich process is the most widely used fluidized bed system for ethylene oxychlorination [12]. Although the process is highly proprietary, a generalized description can be developed from patents and technical articles [11, 13].

Compressed air, HCl from the EDC cracking section, and ethylene are preheated to about 150–170°C and introduced to the bottom of a single-shell fluidized bed reactor. The reactant gases are distributed at the bottom of the reactor and then fluidize the catalyst. Temperature is controlled by internal cooling coils directly submerged in the fluid bed [11]. Because the fluid bed is essentially isothermal, all the reaction occurs at the same temperature. Typically, this temperature is in the range 220–225°C. Pressures are slightly elevated (25–35 psig) to increase the reaction efficiency and to aid the downstream EDC condensation.

As mentioned previously, the process employs a Deacon catalyst with cupric chloride as the catalytic agent. One commercial catalyst manufacturer reportedly makes a catalyst with 10 wt% $CuCl_2$ [14]. B.F. Goodrich has indicated a preferred range of 3.5–7.0 wt% copper, or 7.4–14.8 wt% $CuCl_2$. If the copper level exceeds 12%, reaction rates are not improved and the catalyst has a tendency to cake in the reactor [13].

It is essential for the catalyst support to possess good fluidization properties, which include:

(1)   High adsorptivity for the impregnating salts so that the particles are not sticky,

(2)   Good attrition resistance to reduce fines generation and loss of catalyst, and

(3)   A particle size distribution favorable for fluidization.

High-surface-area (150–250 m^2/g) alumina powder is the preferred support. In U.S. Patent 3,488,398 [13], the following size distribution was given for a representative alumina support:

Diameter (μm)	Weight percent
>80	24
40–80	41
20–40	29
<20	6

Because reaction temperatures are moderate and none of the catalyst experiences extreme temperatures, loss of catalyst activity normally is not a problem. Fresh catalyst can be added to the fluid bed to compensate for catalyst fines lost by attrition or in exchange for catalyst that has lost activity.

Stoichiometric excesses of ethylene and air are used to achieve high HCl conversions. Excesses are usually in the range 2–20% for ethylene and 10–80% for air [5]. Most of the unreacted ethylene passes through the downstream processing without being adsorbed or condensed. High air

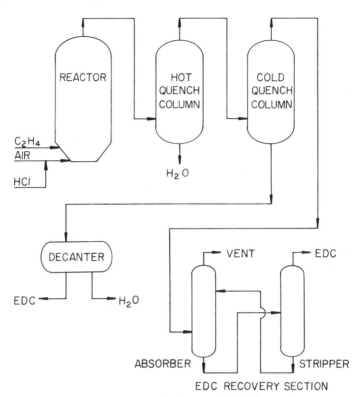

Fig. 2. Fluidized bed ethylene oxychlorination system.

excesses tend to promote ethylene oxidation, which reduces the EDC selectivity from ethylene.

The oxychlorination reaction is very selective. Ethylene dichloride yields of at least 98% from HCl and 96% from ethylene are achievable [15]. Only minor amounts of chlorinated hydrocarbon by-products are produced. These include chloral, 1,1,2-trichloroethane, chloroform, *cis-* and *trans-*1,2-dichloroethylene, and ethyl chloride.

Figure 2 is a simplified flowsheet showing the major processing steps in Goodrich's oxychlorination section [11, 13]. Before leaving the reactor, the product gases pass though a cyclone where catalyst fines are separated from the gas stream and returned to the reaction zone. Next, essentially all the HCl and some of the water is condensed in the hot quench column. The uncondensed gases are directed to the cold quench column where the remaining water and most of the EDC is condensed. The aqueous and organic phases are separated in the decanter. The organic phase is crude, wet EDC, which will be dried and purified before it is cracked.

A secondary recovery system is used to capture the rest of the EDC, which may represent as much as 5% of the total EDC production. Ethylene dichloride is absorbed from the gas stream by solvent extraction using an aromatic solvent. Then the EDC is separated from the solvent in the solvent recovery section. The solvent is recycled to the absorber, and the EDC is added to the wet, crude EDC.

B. PPG FLUIDIZED BED OXYCHLORINATION

Another commercially significant fluidized bed ethylene oxychlorination process is the one developed by PPG. Like the B.F. Goodrich process, the PPG process was commercialized in the early 1960s [16]. The PPG process differs substantially from the Goodrich process; PPG's reactor is multitubed and uses oxygen as the oxidant rather than air. PPG not only has used the process in its own operations but has licensed it to several other companies.

PPG has obtained a large number of patents in the oxychlorination field. Consequently, some variation in PPG-designed plants is expected, depending on when the plant was built and local considerations—such as air pollution restrictions. One early system employed multiple heat exchange tubes contained in a common reactor shell. The reaction heat was removed by a boiling-cooling medium, such as Therminol, that circulated through the tubes and then to an external heat exchanger [17].

Because oxygen is the oxidant, PPG's reactors can operate on a once-through basis or with recycle of the noncondensable off-gases. Recycle operation "allows the oxygen-based process to be operated with an excess of ethylene, thereby enhancing the HCl conversion and EDC quality without sacrificing ethylene yield" [15]. The process vent stream is about a factor of 100 smaller than those with air-based processes. Air pollution and disposal problems are correspondingly reduced.

PPG uses a copper chloride–potassium chloride catalyst on a calcined Fuller's earth carrier, with the trade name Florex [17]. The catalyst contains about 6–12 wt% copper. Most of the particles are in the 30 × 60 mesh (U.S. Sieve) size range because this results in a catalyst with good fluidization properties. If the reactor is operated in a once-through mode, moderate excesses of ethylene and oxygen are used. In the recycle mode of operation, ethylene excesses are typically 30–60% [18]. The higher ethylene partial pressure is advantageous for a selective oxychlorination reaction and improves temperature control. Because the unreacted ethylene is recycled, there is no raw material penalty for using a higher excess. Overall yields are about comparable to those of the B.F. Goodrich process.

C. OTHER FLUIDIZED BED
OXYCHLORINATION PROCESSES

Several other fluidized bed ethylene oxychlorination processes are important commercially. Included in this group are processes developed by Ethyl, Rhone-Poulenc, Monsanto, and Mitsui Toatsu. Ethyl's technology was developed very early and was commercialized by Solvay in the mid-1960s [19]. This technology, which has been jointly licensed with ICI and Solvay, has been installed in 14 plants that produce about 5 billion lb of VCM per year [12]. The fluidizing medium is air, and a single, large fluidization chamber is used. The Mitsui Toatsu process uses oxygen, high ethylene excesses, and recycle of the unreacted ethylene to the primary reactor. These fluidized bed processes will not be described in detail because the B.F. Goodrich and PPG processes are representative of the fluidized bed processes.

IV. Detailed Process Description — Stauffer
Fixed Bed Ethylene Oxychlorination Process

Stauffer's fixed bed ethylene oxychlorination process is the most widely used fixed bed oxychlorination process. Stauffer has licensed its oxychlorination process to over 20 companies worldwide. The second most widely commercialized fixed bed oxychlorination process is the Dow process. Dow started using oxychlorination to make EDC at Freeport, Texas, and Plaquemine, Louisiana, in the early 1960s [19]. Since then, Dow and its foreign affiliates have installed several ethylene oxychlorination units. Technical information on the Dow process is very sketchy because very little information has been published. None of the other fixed bed processes (see Table I) are particularly significant in terms of worldwide EDC production.

The remaining sections of this chapter will focus on ethylene oxychlorination in fixed bed systems. The information is largely representative of the Stauffer process because of its industrial significance, the unavailability of information on the Dow process, and the experience of the authors.

A. REACTOR DESCRIPTION

1. Equipment

Fixed bed ethylene oxychlorination units are normally of multistage design. This provides more flexibility for control of the highly exothermic oxychlorination reaction. Figure 3 represents a three-stage oxychlorination section followed by an ethylene recovery reactor. The reactors are connected in series, with the discharge of one reactor being fed to the next reactor. Air is fed to each reactor. The reactors are constructed like large tubular heat exchangers; see Fig. 4. The reactors contain many small, vertical tubes welded to a tube sheet at the top and bottom. The reactor tubes, which are approximately 1 in. in diameter, contain the catalyst. The tube diameter is selected so that reaction temperatures will not exceed the level where catalyst damage occurs. In larger-diameter tubes, heat transfer between the center of the tube and the tube walls is lower, resulting in higher catalyst temperatures.

The outer jacket of the reactor is provided with openings near the top and bottom to allow a coolant to flood the outer surface of each reactor tube. The most commonly used coolant is water. The reaction heat is transferred to the water, producing steam that can be utilized in other parts of the process or complex.

Most fixed bed oxychlorination processes use nickel alloy for the reactor tubes. Alloy 201 is preferred because alloy 200 tubes are subject to inter-granular embrittlement at sites of localized high hot spot temperatures [20]. Alloys such as stainless steel or those containing high percentages of iron are normally not used because of potential corrosion problems. The tube sheet and reactor heads are clad with nickel on steel. The reactor shell, which is exposed to water and steam, is made of carbon steel. The piping connecting

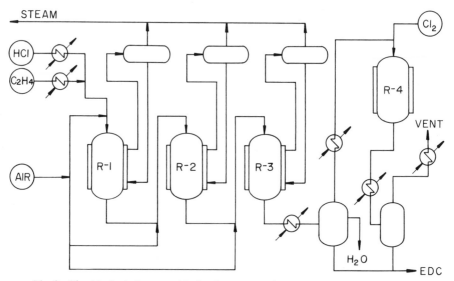

Fig. 3. Fixed bed ethylene oxychlorination system with an ethylene recovery reactor.

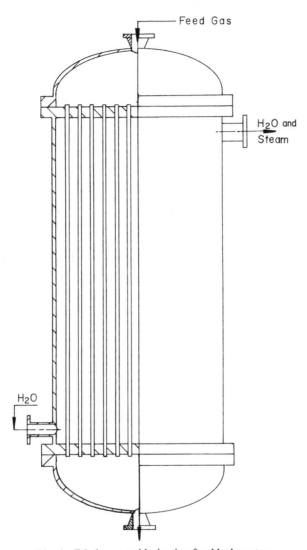

Fig. 4. Ethylene oxychlorination fixed bed reactor.

adjacent reactors is normally nickel or nickel-clad steel. The catalyst is supported at the bottom of the reactor. At the top of the reactor, a perforated nickel plate holds the catalyst in place.

A small number of reactor tubes are equipped with thermowells. The thermowells consist of $\frac{1}{4}$-in. nickel tubes that are placed in the center of the reactor tubes and service the entire length of the tubes. The thermowells are sealed at the lower end and are guided through the top head of the reactor. Temperatures are measured with either a single moving thermocouple or a stationary bundle of thermocouples inserted in the thermowell. The stationary bundle design is used most often because it requires less maintenance. A bundle has several thermocouples with the measurement ends spaced at even intervals. The depth of each bundle varies from one thermowell to the next to obtain composite temperature measurements at smaller intervals. These temperatures are plotted against reactor tube length to produce a temperature profile of the catalyst in the reactor. The measurements are used to establish the maximum reactor temperature, or hot spot temperature, and its location. Based on the temperature profiles, process variables can be adjusted to control the oxychlorination reaction.

Temperatures measured in tubes containing thermowells are somewhat different than temperatures in tubes without thermowells. The presence of the thermowell reduces the packing density of the catalyst and the cross-sectional area available for flow. Hence, the superficial gas velocity in a thermowell tube is greater than that in a nonthermowell tube. Both the higher superficial gas velocity and the less dense catalyst packing favor cooler temperatures in the thermowell tubes. Nevertheless, the measured temperatures are used to adjust the operating parameters.

Other equipment is normally included to ensure homogeneous gas mixtures, to provide diagnostic measures of catalyst condition, and for safety. Each reactor is equipped with a mixer to intimately mix the three reactant feed gases before they enter the reactor tubes. The mixers also help avoid dangerous pockets of explosive mixtures. If an explosive mixture should occur, each reactor is protected against damage from explosion by rupture disks. The diagnostic items include differential pressure transmitters which indicate the pressure drop across each reactor. Increased pressure drop is one indication that the catalyst is physically aging. Thermocouples placed at the exit of each reactor provide an indication of catalyst activity. When the activity drops to a point where the reaction is not complete, the exit temperature from the catalyst bed will be higher because the reaction is still occurring. A skilled operator interprets the higher exit temperature as a decline in catalyst activity and adjusts the operating variables to maintain good conversions.

The gases leaving the final oxychlorination reactor are directed through a water-cooled, corrosion-resistant condenser for the liquid product. The liquid products are received in a vessel which also serves as a phase separator. The uncondensed gas is reheated above its dew point and fed to the ethylene recovery system. The stream from the recovery reactor is passed through a refrigerated condenser to condense as much liquid product as possible before sending the vent gas to downstream processing. The reactor pressure control valve is normally placed after the refrigerated condenser to minimize corrosion.

2. Feed Gases

The ethylene oxychlorination feed gases are hydrogen chloride, ethylene, and air (or oxygen). In most plants all the hydrogen chloride and ethylene are fed to the first stage. Some plants, however, split the HCl stream between the first and second stages to reduce the pressure drop. The amount of air fed to each stage is governed by reaction temperature, oxychlorination efficiency, and/or the explosive envelope. The gases must be preheated before being fed to the reactors, especially for the first stage. The temperature in the reactor head of the first stage must be maintained above the dew point to protect against possible corrosion.

The ethylene and air are fed under ratio control. A ratio controller automatically changes the ethylene and air flows to maintain preselected ratios to the HCl flow. Thus the HCl flow is allowed to vary depending on the pyrolysis production rate, and the ratio controllers change the ethylene and air rates to match the HCl flow.

In a balanced plant, all the HCl generated in the pyrolysis section is fed to oxychlorination. One mole of HCl and 1 mol of vinyl chloride are made from each mole of EDC that is cracked. The stream exiting the pyrolysis unit has HCl, vinyl chloride, uncracked EDC, and minor amounts of by-products, which either were made during cracking or were fed to the cracking unit and emerged unchanged. Hydrogen chloride is separated from the other components in a fractionation column and is then sent to the oxychlorination unit. To minimize fluctuating hydrogen chloride flows to oxychlorination, most plants can send part of their hydrogen chloride production to a liquefaction unit or to other processes.

The hydrogen chloride feed to oxychlorination often contains small amounts of impurities, especially vinyl chloride and acetylene. These impurities increase by-product formation in the oxychlorination reactors. Vinyl chloride is a precursor of 1,1,2-trichloroethane, whereas acetylene leads to several impurities including dichloroethylene, trichloroethylene, and tetrachloroethane. In addition to the increase in by-product formation, some impurities may cause pressure drop problems in the first oxychlorination stage.

The amount of vinyl chloride in the HCl is normally a function of the efficiency of the HCl fractionation column in the VCM purification section of the plant. Some plants with highly efficient columns have very little vinyl chloride in the HCl, whereas others with poor column design have large amounts. The acetylene content, in contrast, is not a function of column efficiency. Although only a small amount of acetylene is formed in the cracking process, almost all of it finds its way to oxychlorination.

Ethylene is usually obtained from a nearby refinery and/or is shipped by rail car or barge from other sites. The ethylene is fractionated in a cryogenic unit to remove impurities such as ethane and methane. Special precautions must be taken with ethylene received by rail car or barge. Often the shipped material contains small amounts of lubricating oil which originated from oil-lubricated compressors used to fill the shipping vessels. Because this oil can cause pressure drop problems, the ethylene is often sent through oil removal units before being fed to oxychlorination.

Oxygen for oxychlorination is provided by compressed air or by pure oxygen. When compressed air is used, the air must be cleaned of oil which may be imparted by the air compressor. Oil in the air has the same effect as oil in the ethylene, namely, increased pressure drop problems. The air compressor intake is positioned so that the air is free from contaminants.

3. Reactor Operation

To control the highly exothermic oxychlorination reaction, the reaction in the first few stages is limited by restricting the air feed. It is common practice to feed all the HCl and ethylene to the first stage and to apportion the air between the stages to maintain reasonable reaction temperatures. The gases exiting the first stage are directed to the inlet of the second stage where they are mixed with the fresh air feed to the second reactor. This scheme is repeated for each of the remaining reactor stages.

HCl conversions by the end of the final stage are normally about 99%. To obtain these high HCl conversions, most plants are operated with stoichiometric excesses of air and ethylene. The stream exiting the final oxychlorination stage is cooled to condense product EDC and water. The vent stream containing unreacted ethylene is reheated and fed to the ethylene recovery reactor. Most recovery reactors are catalytic chlorination units in which chlorine converts the unreacted ethylene to additional product. The exit stream from the recovery reactor passes through a refrigerated condenser to remove as much of the chlorinated hydrocarbons as possible. This gas stream consists mainly of nitrogen (85–90%) plus small amounts of carbon monoxide, carbon dioxide, unreacted oxygen, argon, ethylene, EDC, ethyl chloride, and water. This stream is often incinerated to comply with local air pollution requirements.

The liquid products from oxychlorination and ethylene recovery are neutralized by caustic washing. After phase separation, the product EDC is purified for the pyrolysis unit. Ethylene dichloride purification is normally accomplished by a two-stage distillation operation. The first distillation removes residual water and light ends such as vinyl chloride and ethyl chloride. The second distillation separates product EDC from heavy boilers and is often referred to as a heavy ends distillation. The principal component of the heavy ends is 1,1,2-trichloroethane.

B.   CATALYSTS

1.   Catalyst Base

A solid porous material is normally used to support the catalytically active ingredients. Materials such as alumina, silica–alumina, diatomaceous earth, and activated charcoal are mentioned in patent literature as catalyst supports. Activated alumina in one of its many forms is the most often preferred catalyst support [21]. The alumina can be formed into many shapes. Three of the most common are cylindrical tablets, extrudates, and spheres. Cylindrical tablets are made by physically pressing alumina powder in a pelletizing machine. The tablet density can be adjusted slightly by varying the force used to form the tablets. A small amount of graphite or other lubricant is often used to prevent the tablets from sticking to the pelletizing machine. Extrudates are formed by forcing an alumina paste through a metal sheet containing appropriately sized holes in much the same way that spaghetti is made. The alumina strands are cut to the desired length by a spinning blade. The extrudate cylinders have rough ends in contrast to the smooth surfaces of tablets. Extrudates have a tendency to be softer than tablets and may break more easily.

Many methods exist for making alumina spheres. One method is to roll or dip a hard material, referred to as a seed, in an alumina formulation similar in consistency to paint. The seed is allowed to dry, and the process is repeated until the sphere is the desired size. This process produces spheres of concentric layers. A second method utilizes a tilted spinning plate. A thin alumina paste is slowly dropped on the spinning plate. The alumina gathers and adheres to itself at the center of the plate until the weight of the sphere is large enough to slip off the plate. The size of the sphere can be controlled by the angle and speed at which the plate spins.

After the alumina is formed, the particles are calcined at elevated temperatures to remove moisture and bond the material so that the particles become very cohesive and hard.

Most catalyst manufacturers specify that the alumina support should have certain physical properties. Surface area and pore volume characteristics are among the most important. Calcined alumina catalyst supports normally have surface areas in the range $200–400$ m^2/g. A certain proportion of the surface area must be present in pores that allow easy access to the reactant gas molecules. Pores with diameters between 80 and 600 Å appear to be important [21]. The catalyst carrier must have good attrition hardness so that the resulting catalyst will be able to withstand impregnation, drying, shipping, loading into the reactor tubes, and reaction service. Catalyst size is also important because it affects the packing density in the reactor tubes, hence the reactivity per unit volume of reactor. In addition to physical properties, the catalyst carrier is normally analyzed for several chemical impurities. These include iron, which promotes by-product formation, silica, and sulfur.

2. Catalyst Impregnation

Although copper chloride is invariably the active constituent of oxychlorination catalysts, other ingredients are used to give the catalyst specific properties. The most commonly used additive is potassium chloride, which suppresses the formation of by-product ethyl chloride and reduces the volatility of coppper chloride. The amount of potassium chloride is normally kept low because it tends to decrease the activity of the catalyst for the primary reaction. Potassium chloride forms a low temperature (150°C) eutectic with copper chloride. Thus, at least some of the active ingredients exist in a liquid state on the catalyst surface during the reaction.

The catalyst also may contain other alkali metal chlorides, rare earth metal chlorides, and metallic compounds which assist in promoting the desired reaction and/or inhibiting side reactions [21].

The catalyst is prepared by impregnation, which is accomplished by immersing the alumina into a solution containing the proper amount of chemicals or by spray application in a tumbling bed. After impregnation, the catalyst is dried to remove water.

3. Catalyst Loading — Diluted Catalyst

The principal concern in the operation of a fixed bed oxychlorination process is the control of reaction temperature. High temperatures lead to increased by-product formation, mostly through increased dehydrochlorination of EDC to vinyl chloride followed by additional oxychlorination to give higher chlorinated products. High temperatures also increase the amount of ethylene oxidized to carbon monoxide and carbon dioxide. More important with regard to the catalyst, elevated temperatures may deactivate the catalyst by coking and subsequent powdering of the catalyst particles, and by subliming copper chloride from the catalyst support [12]. These problems are most evident in the vicinity of the hot spot temperatures. Many methods have been proposed to minimize high localized temperatures. Some of these methods include adjusting the reactant ratios, diluting the feed with an inert gas, using a large excess of one or more reactant gases, using tubes of various diameters, and varying the catalyst particle size. Most of these proposed methods for temperature control have some merit. However, the method used most often is catalyst dilution.

The oxychlorination reaction does not produce high temperatures throughout the entire length of the reactor. The reaction proceeds rapidly and vigorously at the inlet of the reactor where the reaction is not limited by any reactant. The temperature in this section may become very high and difficult to control. In a multistage unit, where air or oxygen is split to each stage, oxygen is the limiting reactant. As the oxygen is consumed, the reaction proceeds less vigorously, resulting in cooler bed temperatures. In the lower sections of each reactor, it is thus advantageous to increase the catalyst strength to promote more reaction.

Catalyst dilution has been used successfully to vary the catalyst strength to accomplish good temperature control. An inert diluent such as fused alumina, fused silica, graphite, or other particles having a very low surface area is used to decrease the strength of the catalyst. Catalyst and diluent particles are mixed in proportions so that the catalyst is very dilute close to the reactor inlet. The catalyst concentration increases in the direction of flow, reaching full strength at the bottom of each reactor. The reactors are thus sectioned into several zones, each packed with a different catalyst concentration. Each reactor is charged in a similar fashion, except that the catalyst concentration is higher in the succeeding reactors. Often, no diluent is used in the final reactor.

An example of a diluted catalyst system for a single reactor system will now be described [22]. The catalyst consists of 8.5 wt% $CuCl_2$ on activated alumina. The reactor is divided into four zones. The zone closest to the reactor inlet contains 7 vol% catalyst and 93 vol% graphite diluent. In the second zone, the catalyst/diluent ratio is 15:85 by volume. The third zone contains a 40:60 mixture, and the fourth zone is 100% catalyst. In another variation [21], a relatively concentrated catalyst zone was placed at the very inlet of the first reactor of a three-reactor system to help initiate the reaction. This concentrated zone was followed by a much less concentrated zone to control the reaction temperature. The second reactor had three zones of equal length. The first zone contained a 40:60 mixture of catalyst to diluent, the second zone could be 65:35, and the final zone contained 100% catalyst. The third reactor could have either all 100% catalyst or a less concentrated catalyst in the top section of the reactor.

4. Staged Catalyst

Diluted catalyst systems have several disadvantages. The first is that several different catalyst – diluent mixtures must be formulated. The mixing must be performed carefully to ensure a uniform mixture of catalyst and diluent. However, the catalyst will dust badly if mixed for too long a period of time. Consequently, nonuniform mixtures might be produced. High localized temperatures could result in sections of the reactor where the catalyst is not well mixed. Another disadvantage is the tendency of the catalyst to segregate from the diluent when charging the tubes. This is especially likely if there is a large density difference between the catalyst and the diluent. Another disadvantage is the buildup of pressure drop across the reactor, which results if the shapes of the diluent and catalyst are such that they pack very tightly. This frequently occurs when cylindrical, tableted catalysts are diluted with irregularly shaped fused silica. To compound the problem, catalyst dust that forms may pack between the crevices of adjacent particles, increasing the pressure drop further.

A better catalyst system would be one that either requires no dilution or can be diluted with a diluent of the same shape and density as the catalyst. A catalyst needing no dilution at all is preferred because it is difficult to find a diluent having the same shape and density as the catalyst. The shape of the catalyst (and diluent) should be such that dust is not trapped in the catalyst bed.

Several commercial catalysts have been introduced that eliminate or drastically reduce the need for a diluent. The activity of the catalyst is reduced by loading less of the active ingredient onto the catalyst support, and/or changing the surface area of the support. In some cases, the ratio of copper to potassium changes with catalyst strength. The catalysts are normally offered in three or four activity levels. The strongest catalyst usually has about 15-20 wt% copper chloride and less than 5% potassium chloride. Several manufacturers provide this type of catalyst, although the catalyst shape and support material differ for each manufacturer. The most widely used catalysts are supplied by Stauffer Chemical Company and Badische Anilin- & Soda-Fabrik (BASF). The Stauffer catalyst, a sphere with a size distribution of 3- to 6-mesh (Tyler) sieve, is offered in three impregnation levels. The catalysts are formulated so that no diluent is used in the main catalyst bed. Thus there is no mixing required or particle segregation during loading. The number of zones per reactor for "staged" catalysts is normally less than the number of zones in a diluted catalyst system. Only two or three zones are needed in each of the first two reactors. Thus the number of catalyst formulations for the entire oxychlorination unit might be three or four.

An example of this new type of catalyst and a suggested loading is described for the Stauffer catalyst [21]. Three catalysts were used: A, B, and C. Catalyst A was the weakest of the three, and catalyst C the strongest. Recommended levels of copper chloride and potassium chloride for these catalysts are

Catalyst	A	B	C
CuCl$_2$ (wt%)	6.0	10.0	18.0
KCl (wt%)	3.0	3.0	1.8

The loading pattern for the catalysts was as follows. The first reactor was divided into two zones, the zone closest to the inlet being 60% of the bed and the bottom zone 40%. Catalyst A occupied the inlet zone, and catalyst C was charged into the outlet zone. Reactor 2 was also divided as described for reactor 1. The inlet zone for reactor 2 was loaded with catalyst B and the outlet zone catalyst C. The third reactor was charged with catalyst C throughout.

This oxychlorination catalyst system has performed well commercially. The catalysts are easy to handle while charging and have significantly reduced pressure drop buildup problems. The catalyst is spherical, which gives the bed sufficient voidage to reduce initial pressure drops and to prevent the trapping of dust. This catalyst represents a significant improvement, since the most frequently cited reason for recharging a reactor is a high pressure drop. This improved catalyst has a very long service life. In addition to the improved pressure drop performance, the new catalyst has the capacity to operate at elevated throughput rates while maintaining high HCl conversions. The new staged catalysts generally give higher yields and selectivities than the old diluted systems.

C.   PROCESS VARIABLES

Adjustment of process variables allows a plant operator to control the temperature of the highly exothermic oxychlorination reaction and to optimize yields and selectivities. The effects of most process variables are similar for air- and oxygen-based systems. Process variables that apply only to oxygen-based systems will be discussed in Section D. The following list contains a number of variables that can either be selected or varied to control the performance of a fixed bed oxychlorination unit:

(1)   catalyst selection,
(2)   method of catalyst charging,
(3)   process pressure,
(4)   reactor jacket pressure,
(5)   air distribution,

(6)   HCl split,
(7)   reactor throughput rate,
(8)   excess air and ethylene, and
(9)   preheating of feed gases.

Various combinations of the variables may give the desired reactor performance. Although it is recognized that these variables are interrelated, their effect on oxychlorination will be discussed separately.

1.   Catalyst Selection

Selecting a catalyst system broadly defines catalyst life, optimum throughput rate, temperature behavior, initial pressure drop, pressure drop buildup rate, and maximum yield and selectivity. Although system performance can be modified significantly by varying the process variables, the selection of a catalyst defines the basic performance.

In this paper, catalyst life is defined simply as the length of time the catalyst stays in service. Fixed bed catalysts generally are replaced because of either a high pressure drop or a loss of activity. Catalyst life differs for each of the reactors in the oxychlorination process since process conditions and catalyst compositions differ. The first reactor, which exposes the catalyst to the harshest reaction conditions, has the shortest catalyst life. Catalyst service in the first reactor is about 1 yr at design throughput rates. Catalysts in the other two reactors have service periods one and a half to three times longer than the first reactor. Catalyst life can be influenced by plant parameters such as feed impurities and operating conditions. Although there are differences in the life expectancy of various catalysts, these differences are dependent on the particular plant. Thus when a plant selects a catalyst, their expectations must take into account their previous experience.

Catalysts differ markedly in their ability to perform at elevated throughput rates. Figure 5 presents a plot of HCl conversion versus throughput rate for two catalysts to illustrate this sensitivity to throughput. The diluted catalyst system rapidly loses its ability to convert a satisfactory amount of HCl at high throughput rates. The staged catalyst maintains high HCl conversions at elevated throughput rates.

The selection of a catalyst determines the initial pressure drop and, to a certain extent, the rate of pressure drop buildup. A spherical catalyst, for example, would have a lower initial pressure drop than a tableted catalyst of comparable particle size. This lower pressure drop is important during the catalyst break-in period. It gives the operator more freedom to adjust the system pressure for controlling the reaction temperature of the highly active fresh catalyst.

In addition to the differences among the various catalyst types, yields and

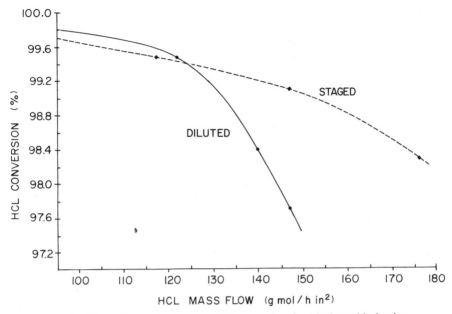

Fig. 5.   Effect of throughput rate on HCl conversion, fixed bed oxychlorination.

selectivities are strongly influenced by process variables and design differences from plant to plant. The number of reactor stages, the air or oxygen distribution, process temperature and pressure, and catalyst age are all important in determining yields and selectivities. Generally, catalysts designed to be used without diluents produce slightly higher yields and selectivities.

2.   Catalyst Charging

Although the catalyst charging operation is not a process variable, it may affect oxychlorination process performance, especially for diluted catalyst systems. These catalyst systems are designed to provide good reaction temperature control, high HCl conversions, and good EDC selectivities. To obtain the desired performance, specific catalyst/diluent ratios must be present in each reactor zone. The catalyst and diluent must be well mixed before charging and then carefully charged to the reactor. If the catalyst is charged too rapidly, catalyst may bridge across some of the reactor tubes. Bridging can be minimized by frequently measuring the catalyst depth to test if it corresponds to the amount of catalyst added. Reduced performance can result from both catalyst bridging and poor catalyst–diluent mixing causing a localized catalyst concentration different than that specified.

Undiluted catalyst systems have fewer charging problems since the mixing step is eliminated. If bridging is avoided, the catalyst should perform as designed. Satisfactory catalyst loading has been obtained both manually and with mechanical chargers.

3. Process Pressure

The oxychlorination process is normally conducted under 4–6 atm of pressure. The process pressure has a direct effect on the volumetric flow rate of the reactants and products in the reactors. Conducting the reaction at high pressures decreases the gas volume, resulting in a lower superficial velocity. The cooling effect on the catalyst particles is reduced. Decreasing the process pressure has the opposite effect: The gas volume is increased, the superficial velocity is higher, and the catalyst in the hot spot zone is cooled more effectively. To visualize the effect, consider a reactor tube in which the reaction has initiated quickly, has a fairly low maximum temperature of 285°C at 40 in. and cools to 220°C at 78 in. There is a secondary hot spot of 241°C in the more active catalyst zone at 105 in. See Fig. 6, line B. An increase in superficial velocity, which can be obtained by decreasing the process pressure, will shift the reaction initiation point slightly downward into the bed. The temperature will rise to the first hot spot location at a

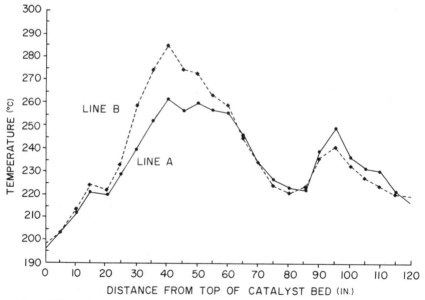

Fig. 6. Effect of process pressure on reactor temperature profile. (A) Steady state pressure; (B) increased pressure.

slower rate because the gases are moving more quickly past the catalyst. In this case, temperature reaches only 262°C. The secondary hot spot rises to a higher temperature because there are more reactants available. See Fig. 6, line A. The catalyst at the bottom of the bed is cooled more slowly, since some of the reaction is still occurring at this location. Increasing the process pressure has just the opposite effect. Reaction initiation occurs earlier, the localized hot spot temperature is higher and is closer to the reactor entrance, and the temperature drops more quickly.

Process pressures are adjusted in commercial practice to control the position and temperature of the hot spots, which are measured by the thermowell tubes. The process pressure is normally set high during start-up to initiate the reaction quickly. The pressure is then reduced to control the reaction temperature. The minimum possible pressure is determined by the pressure required for efficient EDC condensation and the pressure needed to transport EDC for further processing. Normally 30–35 psig at the final reactor exit is the minimum pressure required for these operations. The upper limit for the process pressure, measured at the inlet to the first reactor, is determined by the lowest metering pressure of the feed gases. The lowest pressure is often the delivery pressure of the air compressor.

The system pressure is controlled at the exit of the oxychlorination reactors, and the inlet pressure varies to account for the pressure drops across the reactors. The pressure drop across a catalyst bed depends on the condition of the catalyst and the throughput rate. The pressure drop generally increases as the catalyst becomes older.

4. Reactor Jacket Pressure

The heat generated by the reaction is transferred from the catalyst bed through the reactor tube walls to the boiling jacket fluid, which is usually water. The overall driving force for heat transfer is the temperature difference between the catalyst bed and the cooling fluid in the jacket. If the jacket pressure is increased, the boiling point of the fluid increases and the rate of heat transfer decreases. The net effect is that the catalyst temperature increases. Thus jacket pressure is an independent variable that can be adjusted to control the temperature in the reactor.

The effect of jacket pressure on the temperature profile is very similar to that illustrated for the process pressure. When the jacket pressure is increased, the catalyst bed near the reactor inlet becomes hotter because the driving force for heat transfer is decreased. The temperature reaches a new equilibrium at a higher level in the top part of the reactor. Because of the increased temperature in the top section of the reactor, the reaction rate is faster. The oxygen is consumed more quickly, and the catalyst bed temperature starts to cool higher in the reactor because very little additional reaction occurs. In effect, the hot spot shifts toward the reactor inlet. In a typical commercial run, the jacket pressure is normally held low during the first few months of operation when the catalyst has its highest activity. When the hot spot starts migrating down the reactor, the jacket pressure can be increased to reduce or reverse the migration without overheating the catalyst. The jacket pressure is normally at its highest level during the final few months of operation before the catalyst change-out.

Most catalyst systems are designed so that a combination of adjustments in process pressure and jacket pressure control the position(s) of the hot spot(s). This control is particularly important during the first few weeks of operation with a fresh, highly active catalyst. Fresh catalyst tends to run at a relatively high temperature. Proper temperature control maintains the catalyst in good condition and promotes long catalyst life.

5. Air Distribution

There are three considerations that govern the air split to a multistage fixed bed oxychlorination unit. The first is a safety consideration. The method and rates at which the reactant gases are fed are such that explosive gas mixtures are avoided. The safe selection of air splits to the various oxychlorination reactors was aided by data developed initially by the U.S. Bureau of Mines [23]. The Bureau of Mines developed explosion data for ethylene – air and ethylene – oxygen – nitrogen (or CO_2) mixtures. Based on this work and subsequent studies [24], if all the oxychlorination feed were fed to a single reactor, either fixed bed or fluidized bed, the mixture would be in the explosive envelope. Feeding more than 50% of the air to the first reactor, where all the HCl and ethylene are also fed, provides a mixture close to the explosive envelope. Thus operators of fixed bed oxychlorination units pay close attention to the air feed to the first reactor. Instrumentation is designed so that a reduction in a critical feed, such as HCl, will shut down the plant before the resulting feed mixture reaches the explosive envelope.

Another consideration in the selection of the air split is by-product formation. The oxychlorination reaction produces more CO and CO_2 when more air is fed to the latter stages of the system. Thus to obtain the best yields and selectivities, it is best to feed as much air to the front end of the oxychlorination system as safety considerations permit.

The third consideration, which is very important for freshly charged catalyst, is that the air distribution should not produce high temperatures in any reactor. When operating with a freshly charged catalyst, air distributions are selected to control the reaction temperatures rather than to achieve optimum conversions or selectivities. By maintaining low reaction temperatures, catalyst life is extended and coke formation and catalyst dusting can be minimized. This is advantageous because the generation of catalyst fines increases the pressure drop, whereas coking reduces surface area and hence catalyst activity. A large economic benefit results from careful operation during the catalyst break-in period.

When more air is fed to the latter reactors, oxide formation increases. In this case, the partial pressure of air (O_2) in the first reactor is reduced and that of HCl and ethylene correspondingly increased. These partial pressures promote the formation of ethyl chloride at the expense of the oxychlorination of ethylene. An air split that favors the final reactor may cause high catalyst temperatures in this reactor. If the temperatures in the final reactor become high enough, product EDC will start to crack, forming vinyl chloride and HCl. Indications that this is occurring are a lower HCl conversion coupled with a high catalyst temperature.

At the start of a run, when all three reactors are charged with fresh catalyst, the air split is primarily governed by temperature control. Air to the first reactor is increased until the maximum allowable reactor temperature is reached. The air feed to the second reactor is also adjusted so that its temperature is at the proper level. The remaining air is fed to the final reactor. During operation with fresh catalyst, adjustments to process pressure and jacket pressure are also made so that the unit will operate as cool as possible.

As the catalyst ages, the air feed to the first two reactors can be increased periodically to maintain the reaction temperature at its maximum allowable point. This mode of operation, where the air to the first two reactors is periodically increased, normally continues until an air split is reached that results in the best yields and selectivities. This split is then maintained for the remainder of the run. During the latter stages of operation, the process and jacket pressures are adjusted to minimize migration of the hot spot temperature.

Individual reactor performance can be checked periodically by calculating the ratio of the moles of steam generated in the reactor jacket per mole of air fed to the reactor. This ratio is very stable when the reactor is performing satisfactorily. When the catalyst loses activity, the ratio drops. A substantial drop in this ratio indicates the need to replace the catalyst. The steam/air ratio is not very significant in the final reactor since air is not the limiting reactant in this stage.

6.   HCl Split

Air-based oxychlorination units normally do not split their HCl feed. One of the main reasons for this is that a split HCl feed brings the feed gas mixture of the first reactor closer to the explosive envelope. Another disadvantage of splitting HCl is that the reduction in HCl to the first reactor decreases the rate at which heat can be removed. This results in higher reaction temperatures. Thus the splitting of the HCl feed would not be helpful during the break-in period of a fresh catalyst. By contrast, certain advantages can be realized by splitting the HCl feed, especially toward the end of catalyst service. A split HCl feed reduces the pressure drop across the first reactor, and the lower HCl partial pressure reduces ethyl chloride formation.

7. Reactant Throughput Rates

The effect of throughput rate on air-based, fixed bed oxychlorination systems is complex. Although increased rates mean more reaction and more heat generation, higher catalyst temperatures do not necessarily result. In addition to more reaction and heat generation, the superficial velocity of the gases in the reactors is also higher. The higher gas velocities cool the catalyst more effectively and shift the hot spot down the reactor tubes. Adjustment of jacket and process pressures can return the hot spot to its original position. The resulting temperature profile would invariably be broader. Thus more reaction can be handled without experiencing higher hot spot temperatures.

Conversely, a decrease in throughput rate decreases the superficial velocity of the gases. The reduced gas velocities can shift the hot spot temperature toward the reactor entrance and produce a more narrow temperature peak. Again the jacket and process pressures can be adjusted to broaden the reaction zone and reduce the hot spot temperature.

8. Excess Air and Ethylene

The oxychlorination reaction is normally conducted with stoichiometric excesses of both ethylene and air. This is practiced so that the HCl conversion approaches 100%. Under these conditions, the vent gases from oxychlorination will contain essentially no HCl. The small amount of unreacted HCl dissolves in the water formed by the reaction. The unreacted excess air and ethylene can be separated from the liquid products for further downstream processing.

One effect of not having enough excess air and/or ethylene is a reduction in HCl conversion. Plants normally do not operate with low HCl conversions because the water formed in oxychlorination could require caustic neutralization. Operating the plant with too much excess ethylene has very little effect on performance. There is a slight reduction in hot spot temperatures but, unless the excess is very large, this effect is small. Plants, however, do not operate with too much excess ethylene since the unconverted ethylene must be recovered in downstream processing.

Too much excess air has a variety of effects on the process. The extra air produces higher reaction temperatures, especially in the first reactor where air is the limiting reactant. If enough excess air is used, the reaction can be completed in the first two reactors, causing poor utilization of the third stage. Another negative effect of too much excess air is an increase in the oxidation of ethylene forming CO and CO_2.

9. Preheating of Feed Gases

Preheating the feed gases has several effects on the performance of an oxychlorination system. Poor preheating or lack of preheating the gases going to the first reactor could cause corrosion of the reactor head and top tube sheet. This could occur when there is moisture present in the compressed air feed. The moisture will condense and absorb HCl, forming a highly corrosive liquid.

Feed gas preheating is very important for initiation of the reaction. The oxychlorination feed gases require a certain minimum temperature which must be exceeded before the reaction is initiated. This minimum initiation temperature depends on the process pressure and the reactant concentrations. When poor or no preheating is provided, the catalyst in the top of the reactor tubes preheats the reactant gases to the initiation temperature. The reaction is not initiated at the top of the catalyst bed, but at some distance further down the tubes. Thus, the upper sections of the catalyst are not utilized effectively but merely serve to preheat the gases. Because the initial hot spot temperature is lower in the bed, it has a shorter distance to migrate before the catalyst must be changed. In order to utilize the entire catalyst bed effectively, the gases must be preheated so that reaction initiation occurs high in the catalyst bed.

D. OXYGEN-BASED OXYCHLORINATION

1. Process Description

The main difference between air- and oxygen-based oxychlorination systems is the use of molecular oxygen in place of air. The advantage of feeding oxygen is that it eliminates the large volumes of nitrogen that must be vented from air-based systems. The vent stream from an oxygen-based system is 20–100 times smaller than that from an air-based system [25]. This reduction in the vent stream makes destruction of the environmentally objectionable compounds more manageable. The savings in incineration expenses can offset the cost for oxygen [15].

The nitrogen in an air-based system provides two important functions: It dilutes the reactant mixture to avoid the explosive range, and it removes heat from the localized hot spots. In oxygen-based systems, these requirements can be satisfied by using large excesses of ethylene [26]. After passing through the oxychlorination reactors, most of the ethylene is compressed and recycled. Ethylene moderates temperature better than nitrogen because it has a higher heat capacity. Thus if ethylene is substituted mole per mole for the nitrogen, the catalyst temperature is much lower than for air-based oxychlorination. The cooler catalyst temperatures permit the throughput rate to be increased until the temperature or pressure reaches its maximum limit. Capacity increases of up to 100% over air-based systems have been demonstrated in pilot plant tests [25].

Other advantages of oxygen-based oxychlorination over air-based systems are as follows. Catalyst life is longer because of lower operating temperatures. Product yields are higher as a result of reduced by-product formation. By-product formation in oxygen-based systems is about half that experienced in air-based systems. Product losses in oxygen-based systems are reduced because the vent gas stream is much smaller. The vent gas consists of a small purge removed from the recycle loop to prevent buildup of inert gases such as carbon oxides. The purge gas is dried and sent to a liquid phase direct chlorination reactor where the ethylene is converted to EDC. In addition to the recovery of ethylene, the liquid phase absorbs chlorinated hydrocarbons present in low concentrations, thus reducing their emissions.

The principal disadvantage of oxygen-based systems could be the costs for oxygen and for compressing the recycle gas stream. These costs, however, are offset by higher EDC yields and lower capital investment for the oxygen-based systems.

2. Equipment

The equipment for oxygen-based units is similar to that of air-based oxychlorination with a few exceptions. The only major addition is a compressor for recycling the ethylene-rich vent gases. Special piping and gaskets must be used to handle pure oxygen. The pipes must be free of oil and grease, and gaskets must be made of noncombustible materials. Shutdown procedures must be such that all oxygen is swept away by some inert gas to avoid explosive mixtures.

3. Process Variables

The process variables described for air-based systems generally apply to oxygen-based units. In addition to these variables, HCl split, C_2H_4 split, and the amount of recycle play an important part in oxygen-based units. The effect of splitting HCl and/or ethylene is to lower the ethyl chloride formation and total pressure drop.

The recycle flow rate affects the proximity of the feed gas mixtures to the explosive envelope and the catalyst reaction temperature. By selecting the proper feed proportions, the resultant gas mixtures are not allowed to become explosive. Recycle flows above those required to make the gas mixtures safe can be used effectively to control the reaction temperature.

V.   Future Trends

Several economic, demographic, and political factors should influence the technical character of the international VCM industry in the 1980s. The following major trends are expected:

(1)   A strong trend toward large VCM facilities in industrially advanced countries,

(2)   A continued impact of environmental regulations on VCM plant design, and

(3)   Significant VCM capacity being installed in industrially emerging nations.

Polyvinyl chloride should continue to be favorably priced compared with other polymeric materials because only part of the material is derived from petroleum. Consequently, moderate market growth is expected in both industrially advanced and emerging countries.

A significant trend to large-scale VCM manufacturing facilities is already evident in the United States and, to a lesser extent, in Europe and Japan. Several U.S. companies including Dow, B.F. Goodrich, PPG, and Shell have manufacturing complexes capable of producing approximately a billion lb of VCM per year. More than 80% of the VCM capacity in the United States is in plants with capacities greater than 600 million lb/yr. The trend toward large facilities appears to be self-perpetuating, since small plants cannot compete economically. Increased VCM capacity is expected to be from either new, large, integrated complexes or expansions of existing plants. Oxygen-based oxychlorination technology is favored, particularly for expansions of fixed bed systems; switching from air-based to oxygen-based operation can increase production without additional investment for reactors.

Stringent environmental regulations will have an increasing effect on VCM technology in the more advanced industrial nations. Because of low allowable hydrocarbon emissions, incineration of vent gases will be common. This trend is already established in the United States and Europe. Strict environmental regulations favor the use of oxygen-based oxychlorination because large volumes of nitrogen in the vent stream are avoided. Thus, for the more technically advanced nations, both economic and environmental factors indicate an increased use of oxygen and large-scale plants.

Many technically emerging countries are expected to install VCM manufacturing facilities. These countries include major oil producers, which will market VCM or EDC as higher level products than crude oil, and countries that are becoming more industrialized. Some countries, such as Morocco and Portugal, have already built or planned VCM plants primarily for internal consumption. Densely populated countries, such as India and China, will increase their VCM manufacturing capability to provide PVC for their populations. A varied pattern of VCM plant sizes is expected because plant size depends on raw material availability, product need, and the capacity of associated facilities such as chlorine–caustic facilities. Both fixed bed and fluid bed plants are expected in these countries.

References

1. D. P. Keane, R. B. Stobaugh, and P. L. Townsend, *Hydrocarbon Processing,* February, p. 100 (1973).
2. *Chemical Engineering,* November 8, pp. 245–246 (1965).
3. F. A. Lowenheim, and M. K. Moran, "Faith, Keyes, and Clark's Industrial Chemicals" (Fourth edition), p. 868. John Wiley and Sons, New York (1975).
4. L. G. Shelton, D. E. Hamilton, and R. H. Fisackerly, In "Vinyl and Diene Monomers," (E. C. Leonard, ed.). Wiley-Interscience, New York (1971).
5. J. A. Cowfer, D. E. Jablonski, R. M. Kovach, and A. J. Magistro, U.S. Patent No. 4,226,798 (1980).
6. R. T. Foster, U.S. Patent No. 4,172,052 (1979).
7. J. A. Allen, *J. Appl. Chem. (London)* **12,** 406–412 (1962).
8. R. V. Carrubba, Kinetics and mechanism of the oxychlorination of ethylene, pp. 132, 142. Eng. ScD. Thesis, Columbia University (1968).
9. R. P. Arganbright, and W. F. Yates, *J. Org. Chem.* **27,** 1205–1208 (1962).
10. M. L. Spector, H. Heinemann, and K. D. Miller, *Ind. Eng. Chem. Process Des. Dev.* **6,** 327 (1967).
11. W. S. Amato, B. Bandyopadhyay, B. E. Kurtz, and R. H. Fitch, EPA Report 600/2–76–053, p. 5 (1976).
12. R. W. McPherson, C. M. Starks, and G. J. Fryar, *Hydrocarbon Processing,* March, p. 80 (1979).
13. J. W. Harpring, A. E. Van Antwerp, R. F. Sterbenz, and T. L. Kang, U.S. Patent 3,488,398 (1970).

14. C. L. Thomas, "Catalytic Processes and Proven Catalysts," p. 214. Academic Press, New York (1970).
15. W. E. Wimer, and R. E. Feathers, *Hydrocarbon Processing,* March, p. 83 (1976).
16. J. A. Buckley, *Hydrocarbon Processing,* November, p. 215 (1975).
17. R. M. Vancamp, P. S. Minor, and A. P. Muren, U.S. Patent No. 3,679,373 (1972).
18. A. P. Muren, L. W. Piester, and R. M. Vancamp, British Patent No. 1,220,394 (1968).
19. *Chemical Week,* August 22, p. 94 (1964).
20. C. M. Schillmoller, *Hydrocarbon Processing,* March, pp. 89–93 (1979).
21. R. G. Campbell, E. P. Doane, M. H. Heines J. S. Naworski, and H. J. Vogt, U.S. Patent 4,206,180 (1980).
22. S. E. Penner, and E. M. DeForest, U.S. Patent No. 3,184,515 (1962).
23. Bureau of Mines, Limits of flammability of gases and vapors," p. 68–72, Bulletin 503 U.S. Gov. Printing Office, Washington, D.C.(1952).
24. *Combustion and Flame,* No. 10, p. 95–100 (1966).
25. P. Reich, *Hydrocarbon Processing,* March, pp. 85–89 (1976).
26. A. T. Kister, U.S. Patent No. 3,892,816 (1972).
27. H. Heinemann, *Chem Tech,* May, p. 287 (1971).

CHAPTER 10

Methanol Carbonylation to Acetic Acid

R. T. EBY   T. C. SINGLETON

Process Technology Department
Monsanto Company
Texas City, Texas

I. History of Acetic Acid Processes

Throughout civilization acetic acid (CH_3COOH) has played an important role. For many years the only source of this acid was the oxidation of ethanol in fermented liquids. It was not until 1920 in the United States that catalytic production of acetic acid was begun by the oxidation of acetaldehyde in the presence of manganous acetate. From these beginnings, both the need for improved production methods and the development of these methods grew rapidly. In 1960, Badische Anilin- & Soda-Fabrik AG (BASF) introduced the carbonylation of methanol to acetic acid to the list of several oxidation processes currently used. This method employs the bubbling of carbon monoxide through a reacting mixture containing cobalt as a catalyst at 180°C and 3000–10,000 psig. Acetic acid yields of 90% based on methanol and 70% based on CO have been reported by Von Kutepow *et al.* [1]. This was one of the first examples of commercially reacting simple reactants to form the more complex acetic acid molecule. All previous chemistry had involved various levels of oxidation of ethanol, acetaldehyde, or hydrocarbons to produce the acetic acid. One of these plants, licensed by BASF, was

    275

the Borden Company plant in Geismar, Louisiana. It was the only BASF process built in the United States before a new breakthrough occurred.

In the late 1960s the Monsanto Company developed a process for carbonylating methanol in the presence of a rhodium catalyst to produce acetic acid in high yields at low pressures and reasonable temperatures. High selectivity with a minimum number of by-products has made this process the current choice of companies throughout the world. After successful initial batch studies, development of a continuous process was initiated. Close coordination among the development, engineering, and manufacturing groups, resulted in the start-up of a commercial unit just 4 yr after generation of the idea in 1966. Major hurdles overcome during its development involved materials of construction, catalyst losses, catalyst inventory, product purity, and process control. The resulting process proved to be a simple, efficient, computer-controlled unit producing acetic acid that exceeded the quality of USP, food grade, and reagent grade acid and was suitable for all commercial uses. Monsanto was awarded the Kirkpatrick Merit Award in 1971 for this development.

II. Process Description

Acetic acid is produced by continuously reacting methanol and carbon monoxide in a homogeneous catalytic reactor at <200°C and <500 psi pressure, as shown schematically in Fig. 1. The product is glacial acetic acid of 99.9+% purity. Small amounts of propionic acid, hydrogen, methane, and carbon dioxide are produced in the process. Propionic acid is incinerated. Hydrogen, methane, and carbon dioxide are flared.

A. REACTION AREA

Acetic acid is produced in a liquid phase system in a continuously agitated reactor. The reaction is catalyzed by soluble rhodium compounds with an iodine promoter. There are two primary reactions that occur in the acetic acid process that consume reactants. These are carbonylation of methanol with carbon monoxide to form acetic acid, and a water gas shift reaction that forms carbon dioxide and hydrogen from carbon monoxide and water. As in any catalyst and promoter system, the rhodium and iodine compounds participate in exchange reactions but are not consumed. The theoretical reaction rate of the carbonylation reaction has been determined to be a function of at least temperature, promoter, and rhodium concentration. It has been found that the theoretical reaction rate of methanol carbonylation

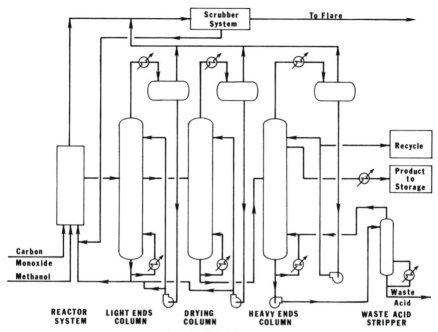

Fig. 1. Monsanto acetic acid process.

increases with (a) increases in temperature, (b) increases in iodine concentration, and (c) increases in rhodium concentration. It has also been found that the reaction rate is independent of the concentration of methanol and carbon monoxide as long as both methanol and carbon monoxide are available for reaction. This condition requires the carbon monoxide to be dissolved in the reaction liquid before it can participate in the reaction. The two conditions that determine the speed with which carbon monoxide goes into solution are the rate of agitation and the partial pressure of carbon monoxide in the gas above the liquid.

An unusual aspect of the acetic acid process is that the reactor is not operated at the maximum theoretical rate because at the maximum theoretical methanol feed rate any slight variation in operating conditions could mean that more methanol would be fed than could be reacted. Such a situation, if continued, would result in upset conditions, making the reaction and purification sections difficult to operate.

B. PURIFICATION AREA

The resulting crude acetic acid is sent from the reaction system to the first column which separates the light ends from a heavy recycle stream. Wet acetic acid is sent as a side stream into the drying column. A water–acetic acid mixture is removed overhead in the drying column and recycled to the reactor to provide part of the cooling for the reaction. Dry acetic acid is sent to the final purification column in which propionic acid is separated as a waste stream from the bottom. Product acetic acid is withdrawn as a side stream, cooled, and sent to storage. The waste stream is concentrated in a small waste acid stripper. The inert gases resulting from the water gas shift reaction are sent to a scrubber system that recovers light ends before the gas reaches the flare.

C. CONTROL

It became obvious early in the development of this process that efficient control of the variables was necessary to ensure high yields and control the effect several recycles have on the process. An efficient supervisory control was developed that ensures methanol yields above 98% and greater than 90% yields for carbon monoxide. The scheme effectively controls the system's heat balance.

In addition to the continuous-process equipment, an auxiliary system operates under batch conditions to prepare the catalyst and promoter and to regenerate the catalyst.

III. Licensing Activities

In 1973, a licensing program began. The first license was granted to the USSR. Since then, a total of nine active licenses have been granted throughout the world, as shown in Table I. Of these, five plants are on stream: USSR,

TABLE I

Acetic Acid Licensees (Monsanto Process)

Company	Country	Capacity (MTA)
Celanese	United States	272,000
USI	United States	272,000
Dainihon	Japan	200,000
Kyodo Sakusan	Japan	200,000
British Petroleum	England	170,000
Techmashimport	USSR	150,000
Rhone-Poulenc	France	225,000
MSK	Yugoslavia	100,000
China Petroleum Development Corporation	Taiwan	80,000

Celanese, USI, Kyodo Sakusan, and Rhone-Poulenc. These plants represent over 90% of the world's new production since 1973. Together with Monsanto's plant, carbonylation of methanol by the Monsanto process represents about 40% of the world's current production. The oxidation processes suffer from high raw material costs and low yields. Replacement of these processes is foreseen in the future, except where the by-products play an important economic role in a specific situation.

IV. Process Chemistry

A. METHANOL CARBONYLATION REACTION

1. *Effects of Reaction Parameters*

These studies have been carried out with both batch and continuous reactor equipment. Any rhodium compound in a solution under reaction conditions forms an active catalyst. The use of many rhodium compounds for this reaction has been reported [2]. Iodine may be charged as hydrogen iodide, methyl iodide, or iodine. All these iodine components are of equivalent activity. Iodine is reduced to hydrogen iodide by this catalyst system. The reaction proceeds at measurable rates at moderate temperature and carbon monoxide partial pressure ($>150°C$ and >100 psi, respectively). Batch reactions are normally carried out in rapidly stirred autoclaves. Carbon monoxide is fed to maintain constant pressure in the reactor as it is consumed in the reaction. Reaction rates in batch reactions may be determined by either the rate of carbon monoxide uptake or by the rate of decrease in active methyl groups in the catalyst solution. An active methyl group is defined as a methyl (CH_3) group that can be converted to an acetyl

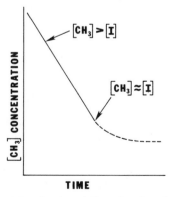

Fig. 2. Rhodium-catalyzed carbonylation of methanol in a batch reaction.

(CH_3CO) group under the conditions of this reaction, e.g., methanol, methyl acetate, methyl iodide, and dimethyl ether. Figure 2 shows a typical rate order for this reaction.

In a batch reaction of this type in which the initial methanol is present in a significant molar excess over the iodide concentration, the carbonylation rate is apparently zero order over a major portion of the reaction. The deviation from zero order occurs at the reaction stage when the molar concentrations of iodide and active methyl groups are approximately equivalent. Under reaction conditions all components of Eqs. (1)–(5) are present in equilibrium concentrations. At normal reaction temperatures ($> 150°C$) these equilibria are established instantaneously.

$$CH_3COOH + CH_3OH \rightleftharpoons CH_3COOCH_3 + H_2O \qquad (1)$$
$$2CH_3OH \rightleftharpoons CH_3OCH_3 + H_2O \qquad (2)$$
$$CH_3OH + HI \rightleftharpoons CH_3I + H_2O \qquad (3)$$
$$CH_3COOCH_3 + HI \rightleftharpoons CH_3I + CH_3COOH \qquad (4)$$
$$CH_3OCH_3 + HI \rightleftharpoons CH_3I + CH_3OH \qquad (5)$$

Dimethyl ether is normally present in the catalyst solution in only a trace amount. Methyl iodide is the equilibrium-favored component in Eqs. (3)–(5). Therefore, when active methyl groups are present in a molar excess over iodine, a major fraction of the iodine exists in equilibrium as methyl iodide. Concentrations of methanol and methyl acetate decrease as these components are converted to acetic acid during the course of the reaction. The methyl iodide concentration remains nearly constant during the zero-order portion of a batch reaction. The rate during this zero-order portion of the reaction is directly proportional to both the rhodium and iodide concentrations over very wide ranges of these two variables.

The carbonylation rate is essentially independent of the initial methanol concentration. The only effect of the initial methanol concentration is a catalyst dilution factor due to an increasing liquid volume during the course of the reaction. This volume change is greater at higher initial methanol concentrations. Water, produced initially by the reactions in Eqs. (1)–(3), is reduced in equilibrium concentration as the reaction proceeds. An initial water concentration over a moderate range has no significant effect on the methanol carbonylation rate. It has been reported that the reaction rate is independent of carbon monoxide partial pressure in the range 200–800 psi [3, 4]. The rhodium–catalyst complex is unstable at low carbon monoxide partial pressure, and therefore the reaction subsides and eventually ceases with decreasing carbon monoxide partial pressure. However, if the minimum carbon monoxide pressure sufficient to sustain an active catalyst is attained, further increases in carbon monoxide partial pressure fail to accelerate the reaction. The minimum carbon monoxide partial pressure is dependent on other reaction parameters, e.g., temperature, concentrations of rhodium and iodine, and vapor pressure of the catalyst solution. This minimum threshold carbon monoxide partial pressure differs with each combination of these reaction parameters.

The anionic iodocarbonyl–rhodate complexes that can exist in acetic acid–water solutions under various conditions under carbon monoxide pressure have been investigated by Denis Forster [5]. The compositions of these anions were assigned on the basis of their infrared spectra and elemental analyses of their quaternary ammonium salts. Samples of catalyst solutions obtained during the zero-order portion of a batch methanol carbonylation reaction were light yellow in color. This color is typical of the d^8 square planar $[Rh(CO)_2I_2]^-$ anion. An *in situ* infrared spectrum of this catalyst solution was obtained by the use of a high-pressure spectrophotometric cell [6]. The spectrum showed bands at 1994 and 2064 cm^{-1}, typical of the $[Rh(CO)_2I_2]^-$ anion. No infrared bands characteristic of other rhodium complexes were observed in the catalyst solution during the zero-order portion of a batch reaction. Near the end of the reaction, when the excess methanol and methyl acetate have been converted, further carbonylation of the remaining methyl iodide increases the hydrogen iodide concentration, as in Eq. (6).

$$CH_3I + H_2O + CO \rightarrow CH_3COOH + HI \qquad (6)$$

As the hydrogen iodide concentration increases, the reaction subsides and, under most conditions, finally stops completely without the total conversion of methyl iodide. An infrared spectrum of the catalyst solution at the end of the reaction showed only a band at 2090 cm^{-1}, typical of the *trans*-$[Rh(CO)_2I_4]^-$ anion. This effect of acidity on the transition of the rhodium complexes is discussed in Section IV.B.1 on the water gas shift reaction.

Iodide salts of alkali metals are inactive as cocatalysts in the rhodium-catalyzed carbonylation of methanol, even though the $[Rh(CO)_2I_2]^-$ complex is formed in the presence of alkali metal iodides. Iodine remains combined as salts of these basic metals under reaction conditions. No methyl iodide was detected in the catalyst solution of an attempted batch reaction with potassium iodide as the iodine source. Iodide salts of more acidic metals, e.g., iron and nickel, are partially active as cocatalysts. These metal iodides are partially solvolyzed and converted to methyl iodide, as in Eq. (7).

$$FeI_2 + CH_3COOCH_3 \rightleftharpoons FeI(CH_3COO) + CH_3I \qquad (7)$$

Analyses of catalyst solution samples from batch reactions with ferrous or nickelous iodide as the iodine source show the presence of low concentrations of methyl iodide. Carbonylation rates in these cases are lower than in reactions in which the iodine source is hydrogen iodide or methyl iodide.

2. Mechanistic Interpretation

The study of reaction parameters demonstrated that the methanol carbonylation rate was first-order-dependent on rhodium and iodine concentrations and independent of methanol concentration and carbon monoxide partial pressure. Under conditions of maximum rate the $[Rh(CO)_2I_2]^-$ and methyl iodide concentrations were in direct proportion to the total rhodium and iodine concentrations, respectively. On this basis it was suggested by Roth *et al.* [4] that the rate-controlling step involves the oxidative addition of methyl iodide to the monovalent $[Rh(CO)_2I_2]^-$ complex. They suggested that this rate-limiting step is followed by a sequence of more rapid reactions involving rearrangements, addition of carbon monoxide, and finally hydrolysis to produce acetic acid. Additions of covalent compounds to d^8 and d^{10} complexes are well known [7, 8]. Additions of carbon monoxide to metal–carbon or metal–hydride complexes have also been reported [9].

A later investigation by Denis Forster [10–12] involved spectrophotometric studies on the rhodium complexes in this reaction cycle. These studies provided evidence for the structures of these intermediates. On the basis of this study a mechanism consistent with the reaction kinetics was proposed for the rhodium–iodine-catalyzed carbonylation of methanol. Forster's mechanism is presented in Fig. 3.

It was observed that, when a solution of $[Rh(CO)_2I_2]^-$ ions was reacted with methyl iodide at ambient temperature, the infrared bands of the original diiododicarbonylrhodate anion were replaced by bands at 2062 and 1711 cm^{-1} in the product. The 1711-cm^{-1} band is typical of an acetyl

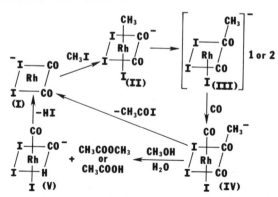

Fig. 3. Proposed mechanism of the rhodium-catalyzed carbonylation of methanol to acetic acid.

frequency. These infrared bands were assigned to intermediate III in the proposed reaction cycle. It is believed that the methyl iodide adduct (intermediate II) exists only in transient form and rapidly rearranges to intermediate III. The elemental analysis of a quaternary amine salt of intermediate III is consistent with an atomic ratio calculated for the salt of intermediate III. The structure of this compound was also confirmed by x-ray diffraction [13]. This x-ray diffraction study indicated that intermediate III existed as a dimer. Vacuum distillation of a tetraphenylarsonium salt of intermediate III produced the $[Rh(CO)_2I_2]^-$ ion, demonstrating the reversibility of these reactions. The treatment of a solution of intermediate III with carbon monoxide at atmospheric pressure and ambient temperature rapidly converted it to a component with CO stretching frequencies at 2141 and 2084 cm^{-1} and an acetyl frequency at 1708 cm^{-1}. This species, which was assigned the structure of intermediate IV, slowly decomposes at room temperature to produce the original complex I of this reaction cycle.

The formation of an acetyl complex (III) in the absence of carbon monoxide and the rapid reaction of intermediate III with carbon monoxide at low pressure are consistent with the lack of rate dependence of the carbonylation reaction on carbon monoxide partial pressure. The conversion of methyl iodide to these intermediates explains the lack of rate dependence on methanol concentration. The first-order rate dependence on both rhodium and iodine suggests that the addition of methyl iodide to the diiododicarbonylrhodate anion (I) is the rate-controlling step in the methanol carbonylation reaction. As mentioned previously, an infrared spectrum of the catalyst solution obtained under actual reaction conditions at elevated temperature showed the presence of only the $[Rh(CO)_2I_2]^-$ anion. The other steps in the reaction cycle after oxidative addition of methyl iodide are apparently very rapid. The other intermediates (II–IV) in the reaction cycle in Fig. 3 are not present in detectable concentrations under steady state conditions. This observation supports the conclusion that the oxidative addition of methyl iodide to the diiododicarbonylrhodate ion is rate-determining in this reaction.

The mechanism proposed by Forster has alternative elimination steps. One of these involves solvolysis of the acetyl complex (IV) to form a hydride intermediate (V), followed by reductive elimination of hydrogen iodide. This is similar to a mechanism suggested for the carboxylation of alkyl chlorides by cobalt catalysts [14]. It was observed by Forster that carbonylation of anhydrous methyl iodide with a tetraphenylarsonium salt of a dihalodicarbonylrhodate catalyst at low temperature and pressure produced detectable amounts of acetyl iodide. Reductive elimination of an acyl halide by carbonylation of a trivalent rhodium phosphine acyl complex has been reported [15]. No oxidative addition of an acetyl halide to a $[Rh(CO)_2X_2]^-$

complex occurred in 24 hr at 50°C. On the basis of these observations, the reductive elimination of acetyl iodide (IV → I) is the favored alternative for the final step of this reaction cycle.

B. COMPETING REACTIONS

The yields of acetic acid, based on both methanol and carbon monoxide, in this process are very high. Minor competing reactions have been observed. These side reactions lead to a very small yield loss. They are discussed in the following sections.

1. Water Gas Shift Reaction

The water gas shift reaction was observed to occur as a side reaction in the rhodium-catalyzed methanol carbonylation reaction in early exploratory and process development studies [4, 16]. In this reaction, carbon monoxide and water are converted to hydrogen and carbon dioxide. This reaction also proceeds at moderate rates in acetic acid solutions in the absence of active methyl groups in this catalyst system under conditions similar to the methanol carbonylation reaction. Workers at the University of Rochester have observed this reaction to proceed at measurable rates in this catalyst system at a low temperature and subatmospheric pressure [17, 18]. The water gas shift reaction has been investigated the most extensively of any of the competing reactions in the rhodium-catalyzed methanol carbonylation process [19, 20].

A. EFFECTS OF REACTION PARAMETERS. Experiments for determining reaction parameter effects were carried out in semibatch reactions by purging carbon monoxide through a catalyst solution of rhodium, hydrogen iodide, water, and acetic acid in a rapidly agitated autoclave at constant pressure and temperature. The water gas rates were determined by the total rate and carbon dioxide content of the off-gas. Mass spectrometric analyses of the off-gas showed the hydrogen and carbon dioxide contents to be equivalent in all cases. These reactions were not carried out in the presence of active methyl groups because of the high volatility of methyl iodide and because of the increase in hydrogen iodide concentration via carbonylation of methyl iodide during the reaction. Steady state conditions could not be attained in a semibatch process in the presence of methyl iodide.

A variable study was carried out to determine the effects of temperature and the concentrations of rhodium, hydrogen iodide, and water on the water gas rate. Four series of runs were carried out. Temperature, rhodium, and hydrogen iodide were maintained at constant levels in each separate group of runs, whereas the water concentration was varied within each group. These experiments demonstrated a complex interaction of hydrogen iodide and water. In each of these series, a peak rate was observed with varying water content when the other variables were constant. The peak rates occurred at higher water levels at the higher hydrogen iodide concentrations.

The presence of sodium iodide also affects the water gas rate. At low water levels the addition of sodium iodide reduces the rate, whereas sodium iodide addition enhances the rate at higher water levels. Water gas rates are enhanced by higher temperature and higher rhodium concentrations. Interactions of carbon monoxide partial pressure with other variables were observed. This effect of carbon monoxide partial pressure is discussed in Section IV.B.1.b, on the mechanism of this reaction.

B.   MECHANISTIC INTERPRETATIONS.   It has been reported that the diiododicarbonylrhodate anion reacts with aqueous hydriodic acid in the absence of carbon monoxide pressure to produce hydrogen, carbon monoxide, and the tetraiodocarbonylrhodate anion, as in Eq. (8) [5, 21].

$$[Rh(CO)_2I_2]^- + 2HI \xrightarrow{\Delta} [Rh(CO)I_4]^- + CO + H_2 \qquad (8)$$

As mentioned previously in Section IV.A.1, an anionic monovalent and four different anionic trivalent rhodium carbonyl iodide complexes have been formed under different conditions (Table II). The structures of these rhodium complexes were assigned on the basis of their infrared spectra and elemental analyses of their tetraalkylammonium salts and, in the case of the trans-$[Rh(CO)_2I_4]^-$ anion, by x-ray crystallography [22].

It is also known that most soluble trivalent rhodium compounds are converted to the monovalent complex $[Rh(CO)_2X_2]^-$ by heating under carbon monoxide pressure in the presence of water and hydrohalic acids with the liberation of carbon dioxide, as in Eq. (9) [23, 24]. It was therefore proposed that the water gas shift reaction in this homogeneous catalyst

TABLE II

Iodocarbonylrhodate Anions and Infrared Bands

Anion	Infrared wave number (cm^{-1})	Oxidation state of rhodium
$[Rh(CO)_2I_2]^-$	1994, 2064	1+
$[Rh(CO)I_4]^-$	2075	3+
cis-$[Rh(CO)_2I_4]^-$	2092, 2122	3+
trans-$[Rh(CO)_2I_4]^-$	2090	3+
$[Rh(CO)I_5]^{2-}$	2047	3+

system occurs by a combination of these oxidation and reduction steps under steady state conditions.

$$Rh^{3+} + 3CO + H_2O + 2X^- \rightarrow [Rh(CO)_2X_2]^- + CO_2 + 2H^+ \qquad (9)$$

An *in situ* infrared spectral study of water gas catalysts at $< 200°C$ and < 500 psig total pressure was performed by obtaining catalyst samples in acetic acid solutions in the high-pressure infrared cell mentioned in Section IV.A.1. This showed that the monovalent diiododicarbonylrhodate anion, $[Rh(CO)_2I_2]^-$, was the predominant rhodium component at low hydrogen iodide and high water concentrations. At high hydrogen iodide and low water levels the rhodium is present primarily as the trivalent *trans*-tetraiododicarbonylrhodate anion, $[Rh(CO)_2I_4]^-$. Similar results were observed in another study with nonanoic acid as solvent.

The tetraiodocarbonylrhodate anion, $[Rh(CO)I_4]^-$, is the favored trivalent rhodium component in coordinating solvents at low carbon monoxide pressure and at high hydrogen iodide and low water levels. In a separate study 0.1 M rhodium solutions were treated with carbon monoxide pressure at 100°C and 45 psig in water–hydrogen iodide–acetic acid mixtures until steady state conditions were obtained. The monocarbonyl tetraiodide anion was the only trivalent rhodium component detected under these conditions. The relative amounts of monovalent and trivalent rhodium carbonyl iodide anions produced under steady state conditions were determined by infrared spectra of the solutions. It was observed that higher concentrations of iodide and lower water levels produced more of the trivalent rhodium complex in the steady state. When the total iodide concentration remains constant, a higher hydrogen iodide fraction leads to a greater proportion of the trivalent complex, whereas a larger fraction of the iodide in the sodium salt form produces more monovalent rhodium in the steady state, as seen in Table III.

TABLE III

Effect of Water and Iodide on the Rhodium Oxidation State
(100°C, 0.1 M Rh, 45 psig total pressure)

Iodide concentration	Total rhodium in Rh(III) state (%)		
	1.65 M H_2O	5.7 M H_2O	7.5 M H_2O
0.3 M HI	11	0	0
0.5 M HI	—	0	0
0.7 M HI	—	36	14
1 M HI	—	100	40
I M NaI	—	0	—
0.3 M HI + 0.7 M NaI	100	10	—
0.5 M HI + 0.5 M NaI	—	50	—

The distribution between the mono- and trivalent rhodium complexes under steady state conditions is dependent on the relative rates of oxidation and reduction steps. As the rate of oxidation of rhodium(I) increases relative to the rate of reduction of rhodium(III), the ratio of rhodium(I) to rhodium(III) should decrease under steady state conditions.

James and Rempel [23, 24] described their studies on the reduction of trivalent rhodium with carbon monoxide and water in aqueous and non-aqueous solutions to produce a monovalent dichlorodicarbonylrhodate anion. The reaction rate is retarded by higher acid concentrations. They suggested that the inverse acid dependence indicates that the rate-determining step is the reaction of carbon monoxide with a hydroxy complex in equilibrium with hydrated rhodium chloride. Hydroxide migration could form a transient carboxylate ligand which decarboxylates to produce a monovalent rhodium ion. Their data did not allow them to determine whether prior coordination of hydroxide ion or water to a trivalent rhodium(III) carbonyl species was necessary for the reaction. A similar mechanism for the water gas reaction catalyzed by rhodium carbonyl iodide complexes has been suggested by Baker et al. [17] and by Cheng et al. [18] at the University of Rochester. Based on several studies (and references therein), the reactions of metal carbonyls with water to generate hydroxycarbonyl species can be interpreted as involving the attack of hydroxide ion on coordinated carbon monoxide without prior coordination of the nucleophile [25].

A similar mechanism consistent with the kinetic studies at Monsanto's laboratories with the acetic acid–iodide–water system is suggested for the step involving reduction of the rhodium(III) anion to the rhodium(I) anion. This reaction should have an inverse rate dependence on acidity. This mechanism is presented in Fig. 4.

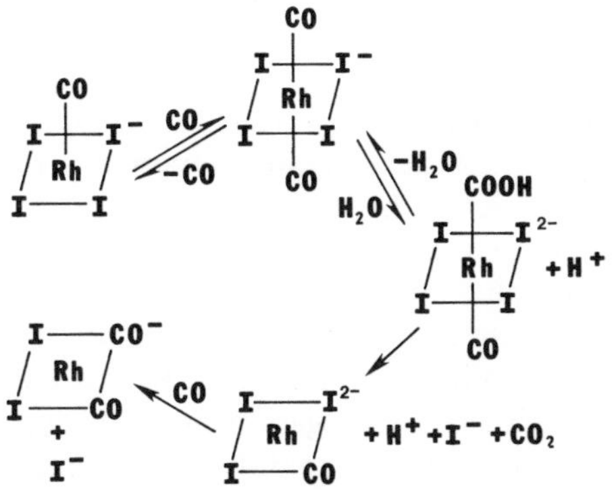

Fig. 4.   Proposed mechanism of the reduction step of the rhodium-catalyzed water gas shift reaction.

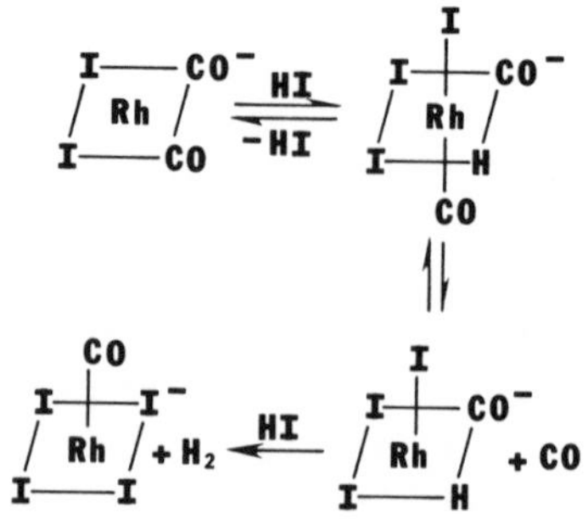

$$\begin{array}{ccc}
\begin{array}{c} \mathrm{I}\!-\!\!-\!\mathrm{CO}^- \\ /\ \mathrm{Rh}\ / \\ \mathrm{I}\!-\!\!-\!\mathrm{CO} \end{array} & \overset{\mathrm{HI}}{\underset{-\mathrm{HI}}{\rightleftharpoons}} & \begin{array}{c} \mathrm{I} \\ \mathrm{I}\!+\!\mathrm{CO}^- \\ /\ \mathrm{Rh}\ / \\ \mathrm{I}\!+\!\mathrm{H} \\ \mathrm{CO} \end{array}
\end{array}$$

$$\Big\updownarrow$$

$$\begin{array}{ccc}
\begin{array}{c} \mathrm{CO} \\ \mathrm{I}\!+\!\mathrm{I}^- \\ /\ \mathrm{Rh}\ /\ +\mathrm{H}_2 \\ \mathrm{I}\!-\!\!-\!\mathrm{I} \end{array} & \overset{\mathrm{HI}}{\longleftarrow} & \begin{array}{c} \mathrm{I} \\ \mathrm{I}\!+\!\mathrm{CO}^- \\ /\ \mathrm{Rh}\ /\ +\mathrm{CO} \\ \mathrm{I}\!-\!\!-\!\mathrm{H} \end{array}
\end{array}$$

Fig. 5. Proposed mechanism of the oxidation step of the rhodium-catalyzed water gas shift reaction.

A suggested mechanism for the oxidation of rhodium(I) to rhodium(III) involves the attack of an acidic species on the $[\mathrm{Rh(CO)_2I_2}]^-$ anion, followed by elimination of carbon monoxide and hydrogen. This oxidation of rhodium(I) should be enhanced by higher acidity and lower carbon monoxide partial pressure. This mechanism is given in Fig. 5.

The structures of the transient complexes in the oxidation and reduction steps are speculative. The experimental data obtained do not definitely establish the nature of the transient complexes.

Under steady state conditions the rates of the two reactions are equivalent. If this proposed mechanism is valid, the overall rate of the water gas reaction can be no greater than the slower of the individual oxidation and reduction steps. At low acidity the oxidation of rhodium(I) should be slow and the reduction of rhodium(III) more rapid. The oxidation step would therefore be rate-controlling. At high acidity the situation should be reversed and the reduction of rhodium(III) should be the slower, rate-controlling step. At low acidity the water gas rate increases with acidity, as the rate of the reduction step decreases until it becomes rate-limiting. The peak water gas rate occurs at this transition point between the rate-controlling steps. The water gas rate then decreases as the acidity is further increased. In this study the neutral Hammett acidity function (H_0) was used to correlate the acidity of these solutions with the water gas rate. Hammett acidity functions were determined for three different hydrogen iodide concentrations at varying water levels in an acetic acid solvent. This relationship of acidity, water, and hydrogen iodide is demonstrated in Fig. 6. Acidity increases with increasing hydrogen iodide concentration. At a constant hydrogen iodide level the acidity goes through a minimum in the 25–35 M water range. However, in the water range used in this investigation of the water gas shift reaction (3–23 M) the acidity increases significantly with decreasing water concentration. The acidity functions were also determined for hydrogen iodide–so-

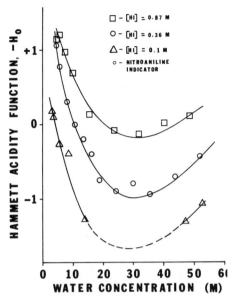

Fig. 6. Effects of water and hydrogen iodide on the Hammett acidity function in acetic acid solutions.

dium iodide–water–acetic acid solutions. The presence of sodium iodide increases the acidity, as demonstrated in Table IV.

Figures 7 and 8 demonstrate a linear relationship of the water gas rate with acidity at varying concentrations of hydrogen iodide, sodium iodide, and water. The peak rate occurs at a Hammett acidity function value $(-H_0)$ of about 0.2 to 0.4. A comparison of rates at two rhodium concentrations shows the rate to be approximately first order in rhodium concentration. A comparison of rates at two temperatures shows the rate to be more temperature sensitive at low acidity, where the rhodium oxidation step is rate-con-

TABLE IV

Effects of Iodide and Water on Acidity

	Hammett acidity function, $-H_0$		
Iodide concentration	7 M H$_2$O	15 M H$_2$O	23 M H$_2$O
0.3 M HI	0.30	−0.56	−0.92
0.3 M HI + O.6 M NaI	0.60	−0.02	−0.65
0.9 M HI	0.88	0.27	−0.03

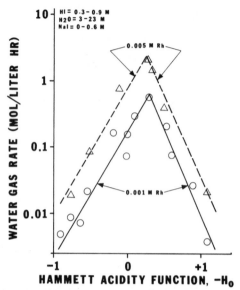

Fig. 7. Effects of acidity and rhodium concentration on the water gas rate in acetic acid solutions.

trolling. This indicates significant differences in activation energies for the oxidation and reduction steps of the reaction.

As mentioned previously, *in situ* infrared studies were carried out on these catalyst solutions. Only the $[Rh(CO)_2I_2]^-$ complex was seen in the low-acidity region where it is proposed that the oxidation of monovalent rhodium is rate-limiting. Only the $[Rh(CO)_2I_4]^-$ complex was observed in the high-acidity region where it is proposed that the reduction of trivalent rhodium is rate-limiting. Both the mono- and trivalent rhodium complexes were detected only in the acidity region of the peak rate. Similar results were also observed by the group at the University of Rochester.

Many of these runs were carried out at 200 and 400 psig total pressure. At acidity function levels greater than that at which the peak rate occurs, the water gas rates at 200 psig were generally equivalent to or slightly lower than the rates at 400 psig (Fig. 9). The reverse is observed at lower acidity levels where the oxidation step is rate-controlling. The rate increases significantly at lower pressure in the low-acidity range.

The researchers at the University of Rochester observed the same effects of acidity and iodide on the water gas rate at a low temperature and subatmospheric pressure. However, one discrepancy between the studies at Monsanto and Rochester involves the effect of carbon monoxide pressure.

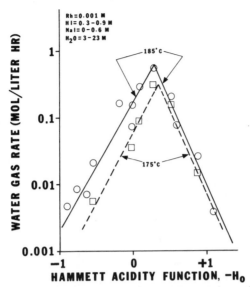

Fig. 8.   Effects of acidity and temperature on the water gas rate in acetic acid solutions.

In the Monsanto study at a higher temperature and pressure the water gas rate was nearly independent of the carbon monoxide pressure at high acidity but was inversely affected by the carbon monoxide pressure at low acidity. The group at Rochester observed the opposite. In their studies at 80 to 100°C and subatmospheric pressure, they observed that the rate was first-order-dependent on carbon monoxide pressure in the high-acidity region but was independent of carbon monoxide pressure at low-acidity levels. At low pressures, it is believed that different steps in the reaction cycle become rate-limiting. In order to accommodate these observations it was proposed by workers at Monsanto that in the rhodium(I) oxidation step carbon monoxide is lost in an equilibrium before hydrogen is liberated [20]. The experimental data do not permit a definite conclusion on whether this loss of carbon monoxide occurs at the rhodium(I) or rhodium(III) stage. The reaction between hydrogen iodide and $[Rh(CO)_2I_2]^-$, when monitored by infrared spectra, gives no evidence of the formation of intermediates. It is believed that these intermediates exist only in transient form. The slight rate enhancement with increasing carbon monoxide pressure, observed when the rhodium(III) reduction step is rate-limiting at the higher temperature and pressure, could be related to an equilibrium between mono- and dicarbonylrhodium(III) species, with the dicarbonyl species being expected to be more reactive toward a nucleophile (H_2O or OH^-) than the monocarbonyl species, in view of the work of Darensbourg and Froelich [25].

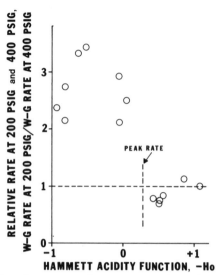

Fig. 9. Effect of pressure on the water gas rate in acetic acid solutions.

C. CORRELATION OF METHANOL CARBONYLATION AND WATER GAS RATES. It was mentioned in Section IV.A.1 that the methanol carbonylation rate is apparently zero order during the portion of a batch run in which the active methyl groups are present in a molar excess over the iodide concentration. During this portion of the reaction the hydrogen iodide concentration is low, because of the favorable equilibrium toward methyl iodide, and the water concentration is high because of esterification. These conditions correspond to the low-acidity region of the water gas rate plot in Figs. 7 and 8 (Section IV.B.1.b). At this acidity level, essentially all the rhodium is therefore present in the $[Rh(CO)_2I_2]^-$ form, which is the active catalyst for methanol carbonylation, and the water gas rate is very low.

At the point in a batch carbonylation reaction where the excess active methyl groups have been converted to acetic acid, carbonylation of methyl iodide begins. Under these conditions hydrogen iodide increases with a simultaneous decrease in water content. As this reaction proceeds, the acidity of the catalyst solution increases. When moderate iodine and water concentrations are present, the fraction of rhodium in the $[Rh(CO)_2I_2]^-$ form decreases and the inactive $[Rh(CO)_2I_4]^-$ content increases as the acidity of the catalyst solution increases. At the stage of the reaction where conditions correspond to the high-acidity region of the water gas rate plot in Figs. 7 and 8, essentially all the catalyst is in the inactive rhodium(III) form and methanol carbonylation ceases. In cases of reactions with low iodine and

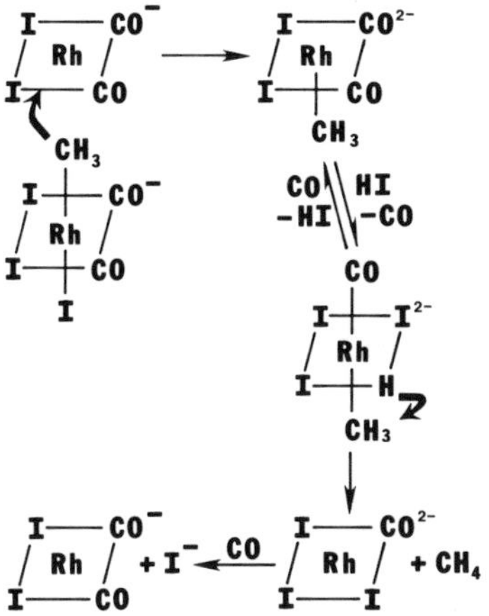

Fig. 10.   Proposed mechanism of methane formation in the rhodium-catalyzed methanol carbonylation reaction.

high water contents the catalyst solution acidity may remain in the region on the low-acidity side of the peak water gas rate plot in Figs. 7 and 8. It is only under these low-acidity conditions that the methanol carbonylation reaction may go essentially to completion.

2.   Methane Formation

Methane is known to be formed in trace amounts in the rhodium-catalyzed methanol carbonylation reaction. A thorough study of this reaction has not been carried out. A speculative mechanism, involving an intermolecular methyl transfer, is proposed in Fig. 10. In this proposed mechanism methane formation is inversely dependent on carbon monoxide partial pressure.

3.   Propionic Acid Formation

Propionic acid is the major liquid by-product in the rhodium-catalyzed methanol carbonylation process. Trace quantities of higher carboxylic acids are also formed. An extensive investigation of higher acids formation has not been carried out. In previously unpublished work it was observed that the amount of propionic acid formed and the concentration of ethyl iodide under steady state were in excess of the amount of ethanol impurity in the methanol fed to the process. It is also known that a higher hydrogen partial pressure in the gas phase in the reactor enhances propionic acid formation. A speculative mechanism involving the reduction of a rhodium–acetyl intermediate complex to a rhodium–ethyl complex is proposed in Fig. 11.

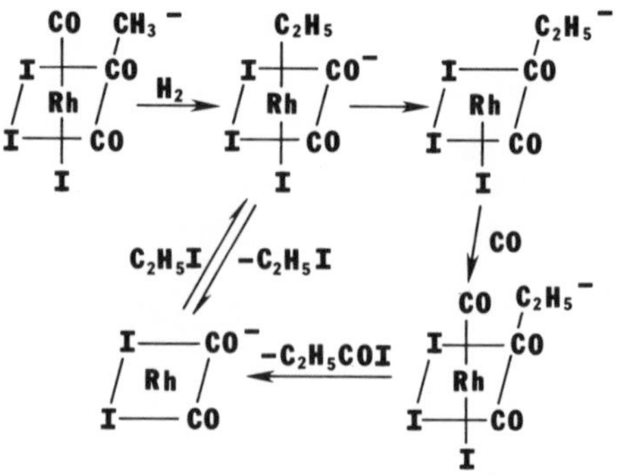

Fig. 11. Proposed mechanism of propionic acid formation in the rhodium-catalyzed methanol carbonation reaction.

Experimental data do not permit a conclusion on whether this reduction of an acetyl group occurs via a reaction with molecular hydrogen or via an intermolecular hydride transfer. In this proposed mechanism the rhodium–ethyl complex, and subsequent propionic acid formation, would be enhanced by higher hydrogen partial pressure or reaction conditions that produce more hydrogen. The rhodium–ethyl complex could undergo further carbon monoxide addition and rearrangement to a propionyl complex. This rearrangement is analogous to that proposed for the mechanism of the methanol carbonylation reaction in Section IV.A.2. Reductive elimination of ethyl iodide from the rhodium–ethyl complex in equilibrium could explain the presence of ethyl iodide in this catalyst solution. Higher carboxylic acids could be formed in the same manner via reduction of the rhodium–propionyl complex and higher homologs. A similar mechanism has been proposed for the synthetic gas homologation reaction catalyzed by a homogeneous ruthenium–iodine catalyst system [26].

Acknowledgments

The authors express their appreciation to Drs. J. H. Craddock, M. J. S. Dewar, L. S. Eubanks, D. Forster, J. Halpern, A. Hershman, D. E. Morris, F. E. Paulik, J. F. Roth, H. B. Tinker, and W. H. Urry for their consultation that was very helpful in understanding the chemistry of this process. A large amount of the spectrophotometric data was obtained by Dr. L. J. Park and Mr. G. L. Roberts. The assistance of Dr. J. F. McGregor and Mr. P. W. Tidwell in the statistical design of the experiments used for the determination of reaction parameter effects is gratefully acknowledged. The drawings reproduced as figures in this chapter were prepared by Mr. J. B. Roubion.

References

1. N. Von Kutepow, W. Himmele, and H. Hohenschutz, *Chem. Ing. Tech.* **37**(4), 383–388 (1965).
2. F. E. Paulik, and J. F. Roth, *Chem. Commun.* 1578 (1968).
3. J. F. Roth, *Platinum Met. Rev.* **19**(1), 12–14 (1975).
4. J. F. Roth, J. H. Craddock, A. Hershman, and F. E. Paulik, *Chem. Technol.* 600–605 (1971).
5. D. Forster, *Inorg, Chem.* **8**, 2556–2558 (1969).
6. D. E. Morris, and H. B. Tinker, *Chem. Technol.* 554–559 (1972).
7. J. P. Collman, *Accounts Chem. Res.* **1**, 136–143 (1968).
8. J. Halpern, *Accounts Chem. Res.* **3**, 386–392 (1970).
9. C. W. Bird, "Transition Metal Intermediates in Organic Synthesis." Logos Press, London (1967).
10. D. Forster, *J. Am. Chem. Soc.* **98**, 846–848 (1976).
11. D. Forster, *Ann. N. Y. Acad. Sci.* **295**, 79–82 (1977).
12. D. Forster, *Adv. Organomet. Chem.* **17**, 255–267 (1979).
13. G. W. Adamson, J. J. Daly, and D. Forster, *J. Organomet. Chem.* **71**, C17–C19 (1974).
14. R. F. Heck, and D. S. Breslow, *J. Am. Chem. Soc.* **85**, 2779–2782 (1963).
15. M. C. Baird, J. T. Mague, J. A. Osborn, and G. Wilkinson, *J. Chem. Soc. A,* 1347–1360 (1967).
16. H. D. Grove, *Hydrocarbon Process.* **51**, 76–78 (1972).
17. E. C. Baker, D. E. Hendriksen, and R. Eisenberg, *J. Am. Chem. Soc,* **102**, 1020–1027 (1980).
18. C. H. Cheng, D. E. Hendriksen, and R. Eisenberg, *J. Am. Chem. Soc.* **99**, 2791–2792 (1977).
19. T. C. Singleton, U. S. Patent 4, 151, 107 (1979).
20. T. C. Singleton, L. J. Park, J. L. Price, and D. Forster, *Prepr. Div. Pet. Chem. Am. Chem. Soc.* **24**, 329–335 (1979).
21. L. M. Vallarino, *Inorg. Chem.* **4**, 161–165 (1965).
22. J. J. Daly, F. Sanz, and D. Forster, *J. Am. Chem. Soc.* **97**, 2551–2553 (1975).
23. B. R. James, and G. L. Rempel, *Chem. Commun.* 158 (1967).
24. B. R. James, and G. L. Rempel, *J. Chem. Soc. A,* 78–84 (1969).
25. D. J. Darensbourg, and J. A. Froelich, *J. Am. Chem. Soc.* **99**, 4726–4729; 5940–5946 (1977).
26. J. F. Knifton, *Prepr. Div. Pet. Chem. Am. Chem. Soc.* **26**, 35–46 (1981).

Bibliography

Additional articles, not cited in the discussion or listed in the Reference Section, describe topics similar to the chemistry of Monsanto's acetic acid process.

D. Brodzki, B. Denise, and G. Pannetier, "Catalytic Properties of Precious Metal Complexes: Carbonylation of Methanol to Form Acetic Acid in the Presence of Iridium Compounds (I)", *J. Mol. Catalysis* **2**, 149–161 (1977).

D. Brodzki, C. Leclere, B. Denise, and G. Pannetier, "Catalytic Properties of Precious Metal Complexes: Carbonylation of Methanol to Acetic Acid by Rhodium Compounds," *Bull. soc. chim. France,* 61–65 (1976).

F. J. Bryant, W. R. Johnson, and T. C. Singleton. "Rhodium Catalyzed Conversion of Methyl Formate to Acetic Acid", *Prepr., Div. Pet. Chem., Am. Chem. Soc.* **18,** 193–195 (1973).

D. Forster. "Relative Stabilities of Some Halide Complexes of Rhodium and Iridium", *Inorg. Chem.* **11,** 1686–1687 (1972).

D. Forster, and G. R. Beck. "Homogeneous Catalytic Decomposition of Formic Acid by Rhodium and Iridium Iodocarbonyl and Hydriodic Acid", *J. Chem. Soc.* **D,** 1072 (1971).

J. Hjortkjaer, and O. R. Jensen. "Rhodium Complex Catalyzed Methanol Carbonylation. Effects of Medium and Various Additives", *Ind. Eng. Chem., Prod. Res. Dev.* **16,** 281–285 (1977).

M. S. Jarrel, and B. C. Gates. "Methanol Carbonylation Catalyzed by a Polymer-Bound Rhodium(I) Complex," *J. Catalysis* **40,** 255–267 (1975).

A. Krzywicki, and G. Pannetier. "Carbonylation of Methanol in the Vapor Phase at Sub-Atmospheric Pressure in the Presence of Rhodium Catalysts. II. Rhodium Chloride Deposited on Alumina", *Bull. soc. chim. France,* 64–68 (1977).

Monsanto Company. "A New Low-Pressure Route to Acetic Acid", *Chem. Eng.* **78,** 64–65 (1971).

G. W. Parshall. "Organometallic Chemistry in Homogeneous Catalysis", *Science* **208,** 1221–1224 (1980).

K. K. Robinson. A. Hershman, J. H. Craddock, and J. F. Roth, "Kinetics of the Catalytic Vapor Phase Carbonylation of Methanol to Acetic Acid", *Symp. Chem. Hydroformylation Relat. React. Proc.,* 111–115 (1972).

R. G. Schultz, and P. D. Montgomery. "Carbonylation of Methanol to Acetic Acid", *J. Catalysis* **13,** 105–106 (1969).

R. G. Schultz, and P. D. Montgomery. "Carbonylation of Methanol to Acetic Acid", *Prepr., Div. Pet. Chem., Am. Chem. Soc.* **17,** B13–B18 (1972).

K. M. Webber, B. C. Gates, and W. Drenth. "Design and Synthesis of a Solid Bifunctional Polymer Catalyst for Methanol Carbonylation", *J. Mol. Catalysis* **3,** 1–9 (1977–1978).

Common Nomenclature and Units in Catalysts

Quantity	Symbol	Common units	SI units	Factor for conversion to SI units
Adsorption				
Volume adsorbed	V	cm³(STP)/g catalyst	m³/kg	1.000E − 03
Monolayer volume	V_m	cm³(STP)/g catalyst	m³/kg	1.000E − 03
Surface area	S_v	m²/cm³ catalyst	m²/m³ catalyst	1.000E + 06
	S_g	m²/g catalyst	m²/kg catalyst	1.000E + 03
Activation energy	E_a	kcal/mol (g)	J/kg mol	4.187E + 06
Catalyst properties				
Particle diameter	d_p	cm	m	1.000E − 02
Particle volume	V_c	cm³	m³	1.000E − 06
Pore volume	V_g	cm³(void)/g catalyst	m³/kg catalyst	1.000E − 03
Pellet density	P	cm³(pellet)/g	m³kg catalyst	1.000E − 03
External area pellet	S_{ext}	m²/g catalyst	m²/kg catalyst	1.000E + 03
Crushing strength	F	lbf/in.²	N/m²	6.895E + 03
		kgf/cm²	N/m²	9.807E + 04
Catalytic reactors				
Reactor length	L	cm	m	1.000E − 02
		ft	m	3.048E − 01
Reactor diameter	D	cm	m	1.000E − 02
		ft	m	3.048E − 01
Reactor volume	V_g	cm³	m³	1.000E − 06
		ft³	m³	2.832E − 02
Fluid velocity	U	cm/sec	m/sec	1.000E − 02
		ft/sec	m/sec	3.048E − 01
Mass velocity	G	g/sec cm²	kg/sec m²	1.000E + 01
Fluid heat capacity	C_p	cal/g mol K	J/kg mol K	4.187E + 03
		btu/lb mol °R	J/kg mol K	4.187E + 03
External pellet area per volume of reactor	a	cm² (pellet)/cm³ (reactor)	m⁻¹	1.000E + 02

Kinetics

Quantity	Symbol	Units	Units	Factor
Total pressure	P_T	atm	N/m²	1.013E + 05
Concentration	C_A	g mol/liter	kg mol/m³	1.000E + 00
		lb mol/ft³	kg mol/m³	1.602E + 01
Quantity of catalyst	W	g	kg	1.000E − 03
		lb	kg	4.536E − 01
Volumetric flow rate	F_0	Liters (reaction conditions)/hr	m³/sec	2.778E − 07
		gal (US)/hr	m³/sec	1.051E − 06
		ft³ (reaction conditions)/hr	m³/sec	7.8658E − 05
Contact (space) time	S	sec	sec	1.000E + 00
Space velocity	$V_g V_0^{-1}$/hr	sec⁻¹	sec	1.000E + 00
Gas hourly space velocity		cm³ (gas)/cm³ (catalyst) hr	m³/m³ sec	2.778E − 04
Liquid hourly space velocity	LHSV	cm³ (liquid)/cm³ (catalyst)/hr	m³/m³ sec	2.778E − 04
Rate of reaction	−r	g mol/cm³ (catalyst) sec	kg/mol/m³ sec	1.000E + 03

Petroleum refining

Quantity	Symbol	Units	Units	Factor
Barrel (42 gal)	Bbl	Barrel	m³	1.590E − 01
Capacity (through-put)	MMBD	Million bbl/day	m³/sec	1.840E + 00
	MMCFD	Million ft³/day	m³/sec	3.278E − 01
	T/D	Tons/day	kg/sec	1.050E − 02

Other

Quantity	Symbol	Units	Units	Factor
Turnover number	N	Molecules per site per second	Molecules per site per second	1.00E + 00
Activation energy	E	kcal/g mol	J/kg mol	4.187E + 06
Adsorption coefficient of substance A	k_A	atm⁻¹	N/m²	9.869E − 06
Heat transfer coefficient	h	cal/sec cm² °C	J/s m² K	4.187E − 04
Gas constant	R	1.982 cal/g mol K	8314.4 J/kg mol K	

Index